AN INTRODUCTION TO
THE APPROXIMATION
OF FUNCTIONS

Theodore *J. Rivlin*

THOMAS J. WATSON RESEARCH CENTER

IBM CORPORATION

DOVER PUBLICATIONS, INC.
NEW YORK

For
Madeline
and
Jean

Published in Canada by General Publishing Company,
Ltd., 30 Lesmill Road, Don Mills, Toronto, Ontario.
Published in the United Kingdom by Constable and Com-
pany, Ltd., 10 Orange Street, London WC2H 7EG.

This Dover edition, first published in 1981, is an unabridged
and corrected republication of the work originally published
by the Blaisdell Publishing Company in Waltham, Massachu-
setts in 1969.

International Standard Book Number: 0-486-64069-8
Library of Congress Catalog Card Number: 80-69675

Manufactured in the United States of America
Dover Publications, Inc.
180 Varick Street
New York, N.Y. 10014

PREFACE

It is by now a commonplace observation that the needs of automatic digital computation have spurred an enormous revival of interest in methods of approximating continuous functions by functions which depend only on a finite number of parameters. The purpose of this book is to provide an introduction to some of the most significant of these methods with particular emphasis on approximation by polynomials, although approximation by piecewise polynomial functions and rational functions is also discussed.

The author views approximation theory as an area of mathematics with important practical applications in computation, and intends to provide here an introduction to the theoretical foundations which underlie many of the algorithms of everyday use. For this reason, for each method of approximation studied at least one algorithm leading to actual numerical approximations is described and, indeed, traced from its theoretical origins to its present formulation. There are, however, no flow charts or actual programs in this book, and the algorithms that are described in detail are intended more to illustrate one possibility than to suggest that they are the "best" available.

Apart from its applications approximation theory is a lively branch of mathematical analysis. The material in this book can be used as additional reading in introductory courses in mathematical analysis as well as numerical analysis. A reader who has studied advanced calculus and the rudiments of linear algebra is amply prepared to understand this book. An effort has been made to avoid more sophisticated prerequisites even at the cost of making the presentation clumsier, as, for example, is the case in Chapter 3 where measure theory, which is the natural language of the material, is not mentioned in order to keep the topic within the grasp of the uninitiated. In keeping with this philosophy and in the interests of good pedagogy the author has not hesitated, at times, to repeat similar arguments, to prove a weaker and then a stronger form of the same theorem, to prove a special case when essentially the same proof gives the general result, to stick to garden variety polynomials, when arbitrary Chebyshev Systems cost very little more and to provide a proof of "something everybody already knows."

Some of the exercises at the end of each chapter verge on the trivial, others are details needed in the text whose inclusion there would needlessly delay the exposition, while still other connected sets of exercises are intended to entice the reader into some interesting side excursions.

References to the literature are given in the usual fashion. A superscripted number in the form [n] refers to item (n) in the notes at the end of the chapter in question.

The author has profited from discussions with many of his colleagues in the Mathematical Sciences Department of the IBM Research Center and welcomes this opportunity to express his thanks to them, as well as to Mrs. Joyce Abish who typed the manuscript in her usual impeccable fashion.

T. J. R.

Bronx, New York

CONTENTS

INTRODUCTION

There are certain generalities about approximation theory that will be useful in our later, more detailed study of specific approximation techniques. The natural setting for these general results is a *normed linear space*. Linear spaces have become familiar objects in mathematics, and so we assume that the reader is familiar with their definition and most elementary properties. We shall take the scalars to be the real numbers unless some other field is specified.

Let V be a linear space. We recall that a *norm* is a function from V into the nonnegative real numbers. This function is written $\|\cdot\|$ and satisfies the following three properties:

(i) $\|v\| \geq 0$ with equality if and only if $v = 0$.

(ii) $\|\lambda v\| = |\lambda| \, \|v\|$ for any scalar λ.   (I.1)

(iii) $\|v + w\| \leq \|v\| + \|w\|$ (the Triangle Inequality).

The norm gives us a notion of *distance* in V. If $w, v \in V$, then the distance from w to v (or v to w) is $\|v - w\|$.

We are now in a position to present the general setting for much of approximation theory. Let W be a subset of V, then, given $v \in V$, the approximation problem, baldly stated, is: Find a $w \in W$ whose distance from v is least; that is, find $w^* \in W$ such that $\|v - w\|$ is least for $w = w^*$. Such a w^* we call a best approximation to v out of W. Problems arise immediately. Is there such a w^*? If there is, is there only one? Since, as we shall see, many of the most widely studied and used methods of approximation are instances of this general approximation problem, we shall save much duplication of effort by obtaining some results in the general situation.

We turn first to the existence question. We have

THEOREM I.1. *If V is a normed linear space and W a finite-dimensional subspace of V, then, given $v \in V$, there exists $w^* \in W$ such that*

$$\|v - w^*\| \leq \|v - w\|$$

for all $w \in W$.

Proof. Since $0 \in W$, it is a competitor for best approximation to v out of W. Its distance from v is $\|v - 0\| = \|v\|$. If $\|v - w\| > \|v\|$, we are, therefore,

1

sure that w cannot possibly be a best approximation to v, and hence we restrict our attention to $w \in W$ which satisfy

$$\|v - w\| \le \|v\| = M.$$

If $\|v - w\| \le M$, then

$$\|w\| = \|-w\| = \|(v - w) + (-v)\| \le \|v - w\| + \|v\| \le 2M.$$

Suppose W is k-dimensional and $w_1, \ldots, w_k$ is a basis for W; then we are trying to prove that the function

$$f(\lambda_1, \ldots, \lambda_k) = \|v - (\lambda_1 w_1 + \lambda_2 w_2 + \cdots + \lambda_k w_k)\| \tag{I.2}$$

takes on a minimum value as the point $(\lambda_1, \ldots, \lambda_k)$ varies in k-space in such a way that

$$\|\lambda_1 w_1 + \cdots + \lambda_k w_k\| \le 2M. \tag{I.3}$$

We want to see what restriction (I.3) implies for the position of the point $(\lambda_1, \ldots, \lambda_k)$. The function

$$g(\lambda_1, \ldots, \lambda_k) = \|\lambda_1 w_1 + \cdots + \lambda_k w_k\| \tag{I.4}$$

is a continuous function of $(\lambda_1, \ldots, \lambda_k)$ (see Exercise I.1) and so assumes its minimum value on the compact set

$$|\lambda_1| + |\lambda_2| + \cdots + |\lambda_k| = 1. \tag{I.5}$$

This minimum value, m, is *positive*. To see this, note that if

$$g(\lambda_1^*, \lambda_2^*, \ldots, \lambda_k^*) = m,$$

then $m \ge 0$. If $m = 0$, then $\|\lambda_1^* w_1 + \cdots + \lambda_k^* w_k\| = 0$, which by (I.1) (i) implies that $\lambda_1^* w_1 + \cdots + \lambda_k^* w_k = 0$. Since $w_1, \ldots, w_k$ are linear independent, we conclude that $\lambda_1^* = \lambda_2^* = \cdots = \lambda_k^* = 0$, thus violating (I.5).

Now if we relax (I.5) and require only that

$$\sum_{i=1}^{k} |\lambda_i| \ne 0,$$

then

$$g\left(\frac{\lambda_1}{\sum_{i=1}^k |\lambda_i|}, \frac{\lambda_2}{\sum_{i=1}^k |\lambda_i|}, \ldots, \frac{\lambda_k}{\sum_{i=1}^k |\lambda_i|}\right) \ge m > 0,$$

and so, in view of (I.1) (ii) we obtain

$$g(\lambda_1, \ldots, \lambda_k) \ge m \sum_{i=1}^{k} |\lambda_i|. \tag{I.6}$$

But (I.6) remains true if $\sum_{i=1}^{k} |\lambda_i| = 0$ and, therefore, is true for all $(\lambda_1, \ldots, \lambda_k)$. Hence, we conclude that (I.3) implies the restriction

$$|\lambda_1| + \cdots + |\lambda_k| \leq \frac{2M}{m}, \tag{I.7}$$

which in turns implies that

$$|\lambda_i| \leq 2\frac{M}{m}, \qquad i = 1, \ldots, k. \tag{I.8}$$

(I.8) defines a hypercube in k-space, which is a compact set, and $f(\lambda_1, \ldots, \lambda_k)$ is a continuous function of $(\lambda_1, \ldots, \lambda_k)$; therefore f assumes its minimum value on (I.8). (As a matter of fact, the point $(\lambda_1, \ldots, \lambda_k)$ at which the minimum value of f is assumed which *might* satisfy (I.3) *must* satisfy (I.3) since all other points have been ruled out as competitors.) The theorem is proved. ∎

Remark. The field of scalars of V could just as well be the complex numbers. The proof is the same.

Let us look at some examples before continuing with the general theory.

Example I.1. The set of functions continuous on a given closed interval $[a, b]$, which we denote by $C[a, b]$, is a linear space. If $f \in C[a, b]$, we can define a norm in $C[a, b]$ by

$$\|f\| = \max_{a \leq x \leq b} |f(x)|. \tag{I.9}$$

The norm is called the *uniform* or *Chebyshev* norm. It is easy to check (and the reader should do so) that (I.9) does, indeed, define a norm. As an example of Theorem I.1 take V to be $C[a, b]$ and let W be the $(n + 1)$-dimensional subspace of $C[a, b]$ spanned by the functions $1, x, \ldots, x^n$. That is, W consists of all polynomials of degree at most n. We call this particular subspace, which plays an important role in our book, P_n. Theorem I.1 now informs us that every continuous function, $f(x)$, on $[a, b]$ has a best approximation out of the polynomials of degree at most n in the uniform norm. That is, given $f \in C[a, b]$, there exists $p^* \in P_n$ such that

$$\max_{a \leq x \leq b} |f(x) - p^*(x)| \leq \max_{a \leq x \leq b} |f(x) - p(x)| \tag{I.10}$$

for all $p \in P_n$.

Notice that the uniform norm singles out x values at which the approximation is worst (that is, where the absolute error is greatest) and assigns as a measure of approximation these worst possibilities. It thus provides absolutely certain bounds on the error at the expense of these bounds having to be large enough to be valid at every point, no matter how exceptional. Another way of expressing (I.10) is

$$\min_{p \in P_n} \max_{a \le x \le b} |f(x) - p(x)| = \max_{a \le x \le b} |f(x) - p^*(x)| \, ;$$

hence the name "min-max" is sometimes used for approximation using the uniform norm.

Example I.2. Another instance in which the uniform norm is widely used is the case of functions defined on a finite point set. Given m distinct real points $x_1 < x_2 < \cdots < x_m$, the set of functions defined on $x_1, \ldots, x_m$ is precisely E_m, the m-dimensional linear space of numerical vectors $f: (f_1, \ldots, f_m)$; f_j may be thought of as the value of $f(x)$ at $x = x_j$, $j = 1, \ldots, m$. The uniform norm on E_m is defined by

$$\|f\| = \max_{i = 1, \ldots, m} |f_i|.$$

(Note that while this defines a norm on E_m, it does not define a norm on, say, $C[x_1, x_m]$ since $\|f\|$ may equal 0 without f being the zero function.) As an application of Theorem I.1 let us take $V = E_m$ and let W be the $(n + 1)$-dimensional subspace of E_m consisting of all vectors $p: (p(x_1), \ldots, p(x_m))$, where $p \in P_n$ and $n < m - 1$. Theorem I.1 tells us that there exists $p^* \in P_n$ such that

$$\max_{i = 1, \ldots, m} |f_i - p^*(x_i)| \le \max_{i = 1, \ldots, m} |f_i - p(x_i)|$$

for all $p \in P_n$.

Example I.3. Instead of the uniform norm in $C[a, b]$, we often consider the norm defined by

$$\|f\| = \left[\int_a^b |f(x)|^p \, dx \right]^{1/p}, \tag{I.11}$$

where p is a real number, $p \ge 1$. Here, also, there is a finite point set analogue. Given m real distinct points $x_1 < x_2 < \cdots < x_m$, we can introduce as norm in E_m

$$\|f\| = \left[\sum_{i=1}^{m} |f_i|^p \right]^{1/p}, \qquad p \ge 1. \tag{I.12}$$

In the case $p = 2$, we recover the usual Euclidean distance in E_m. In both cases of the "p-norm," we conclude from Theorem I.1 that there exists a best approximation to a given f out of P_n.

Example I.4. The requirement that W be *finite*-dimensional in Theorem I.1 is essential. For suppose W is the subspace of $V = C[0, \frac{1}{2}]$ consisting of all polynomials (of any degree). Clearly, W is not finite-dimensional (if it were, what could its dimension be?). We wish to show that $f(x) = 1/(1 - x)$ has no best approximation in the uniform sense on $[0, \frac{1}{2}]$ out of W. Note that, given $\varepsilon > 0$, there exists N such that

$$|f(x) - (1 + x + x^2 + \ldots + x^N)| < \varepsilon, \qquad 0 \le x \le \tfrac{1}{2}.$$

Hence, if there were a best uniform approximation to $f(x)$ out of W, say p^*, it would have to satisfy

$$\|f - p^*\| = 0,$$

which implies that

$$\frac{1}{1 - x} \equiv p^*,$$

an impossibility.

Suppose now that W is a subspace of V and let W^* be the set of best approximations to a given $v \in V$ out of W. (Theorem I.1 gives us a condition under which W^* is not empty.) We wish to prove that W^* is a *convex* set. We recall that a set, S, in a linear space is convex if $s_1, s_2 \in S$ implies that

$$\lambda_1 s_1 + \lambda_2 s_2 \in S$$

if λ_1 and λ_2 are nonnegative and

$$\lambda_1 + \lambda_2 = 1.$$

If S is empty or consists of one point, then it is clearly convex.

THEOREM I.2. *If $v \in V$ and W is a subspace of V, the set of best approximations to v out of W, call it W^*, is convex.*

Proof. If W^* is empty, the theorem is true. Suppose that $w_1^*, w_2^* \in W^*$; then

$$\|v - w_1^*\| = \|v - w_2^*\| = \rho.$$

Suppose $\lambda_1, \lambda_2 \ge 0$ and $\lambda_1 + \lambda_2 = 1$; then

$$\|v - (\lambda_1 w_1^* + \lambda_2 w_2^*)\| = \|\lambda_1(v - w_1^*) + \lambda_2(v - w_2^*)\|$$
$$\le \lambda_1 \|v - w_1^*\| + \lambda_2 \|v - w_2^*\| = (\lambda_1 + \lambda_2)\rho = \rho.$$

Thus, $\lambda_1 w_1^* + \lambda_2 w_2^* \in W^*$, and so W^* is convex. ▮

Theorem I.2 has the consequence that, if there are two distinct best approximations out of W to v, there are infinitely many (in fact, uncountably many) best approximations.

A final general result gives a criterion that insures that, if there is a best approximation, there is only one. The normed linear space V is said to have a *strictly convex norm* if the set B: $\{v/\|v\| \leq 1\}$, called the *unit ball* in V, is *strictly convex* (sometimes called rotund). B is certainly convex. For it to be *strictly convex*, we require that, if $v_1 \neq v_2$, $\|v_1\| = 1$ and $\|v_2\| = 1$, then $\|\lambda_1 v_1 + \lambda_2 v_2\| < 1$ if $\lambda_1, \lambda_2 > 0$ and $\lambda_1 + \lambda_2 = 1$. (That is, the boundary of B contains no open line segment.)

We can now state a uniqueness theorem.

THEOREM I.3. *If V has a strictly convex norm, then a given $v \in V$ has at most one best approximation out of a subspace, W, of V.*

Proof. Suppose w_1^* and w_2^* are two distinct best approximations to v out of W. By Theorem I.2, $(w_1^* + w_2^*)/2$ is also a best approximation to V out of W. Suppose

$$\|v - w_1^*\| = \|v - w_2^*\| = \rho.$$

Put

$$v_1 = (v - w_1^*)/\rho, \qquad v_2 = (v - w_2^*)/\rho.$$

Then $v_1 \neq v_2$, $\|v_1\| = 1$, $\|v_2\| = 1$ and, since the norm in V is *strictly convex*,

$$\left\| \frac{1}{2} v_1 + \frac{1}{2} v_2 \right\| = \left\| \frac{1}{2\rho} (v - w_1^*) + \frac{1}{2\rho} (v - w_2^*) \right\| = \frac{1}{\rho} \left\| v - \frac{w_1^* + w_2^*}{2} \right\| < 1$$

or

$$\left\| v - \frac{w_1^* + w_2^*}{2} \right\| < \rho.$$

This contradicts the definition of ρ; therefore, the theorem is proved. ∎

It now becomes of interest to determine which spaces have strictly convex norms. Let us examine the spaces of Examples I.1–I.3. Turning first to Example I.3 with the norm defined by (I.11), we suppose that $f_1, f_2 \in C[a, b]$, $f_1 \neq f_2$,

$$\|f_1\| = \left[\int_a^b |f_1(x)|^p \, dx \right]^{1/p} = 1, \quad \text{and} \quad \|f_2\| = \left[\int_a^b |f_2(x)|^p \, dx \right]^{1/p} = 1.$$

The triangle inequality implies that $\|\lambda_1 f_1 + \lambda_2 f_2\| \leq 1$ if $\lambda_1, \lambda_2 > 0$, $\lambda_1 + \lambda_2 = 1$. Suppose $\|\lambda_1 f_1 + \lambda_2 f_2\| = 1$ for some $\lambda_1, \lambda_2 > 0$ satisfying $\lambda_1 + \lambda_2 = 1$. We shall show that this is impossible if $p > 1$. To this end, we need the following lemma.

LEMMA I.1. *If $A > 0$ and $B > 0$ and $0 \leq t \leq 1$, then*

$$A^t B^{1-t} \leq tA + (1 - t)B, \tag{I.13}$$

and equality holds in (I.13) only if $t = 0$ or 1, or $A = B$.

Proof. The second derivative of the function $\log(1/x)$ is $1/x^2$ which is positive for positive x. Hence, $\log(1/x)$ is a *convex* function of x for positive

x; that is, the chord joining two points of the curve $y = \log(1/x)$ lies above the curve between the two points. This means that, for $0 \le t \le 1$ and $A, B > 0$,

$$\log \frac{1}{tA + (1-t)B} \le t \log \frac{1}{A} + (1-t) \log \frac{1}{B}$$

with equality possible only if $t = 0, 1$ or $A = B$. The lemma now follows by exponentiation.

We return to the proof of the theorem.

$$|\lambda_1 f_1 + \lambda_2 f_2|^p = |\lambda_1 f_1 + \lambda_2 f_2| \, |\lambda_1 f_1 + \lambda_2 f_2|^{p-1}$$
$$\le \lambda_1 |f_1| \, |\lambda_1 f_1 + \lambda_2 f_2|^{p-1} + \lambda_2 |f_2| \, |\lambda_1 f_1 + \lambda_2 f_2|^{p-1}. \quad (I.14)$$

Equality is possible in (I.14) only if $f_1(x) f_2(x) \ge 0$ for $a \le x \le b$. Let us take

$$A = |f_1|^p, \qquad B = |\lambda_1 f_1 + \lambda_2 f_2|^p,$$

and

$$t = \frac{1}{p}$$

in (I.13); then

$$|f_1| \, |\lambda_1 f_1 + \lambda_2 f_2|^{p-1} \le \frac{|f_1|^p}{p} + \left(1 - \frac{1}{p}\right)|\lambda_1 f_1 + \lambda_2 f_2|^p \quad (I.15)$$

with equality only if $|f_1| = |\lambda_1 f_1 + \lambda_2 f_2|$. Similarly,

$$|f_2| \, |\lambda_1 f_1 + \lambda_2 f_2|^{p-1} \le \frac{|f_2|^p}{p} + \left(1 - \frac{1}{p}\right)|\lambda_1 f_1 + \lambda_2 f_2|^p \quad (I.16)$$

with equality only if $|f_2| = |\lambda_1 f_1 + \lambda_2 f_2|$. Using these inequalities in (I.14) and integrating, we obtain

$$1 = \int_a^b |\lambda_1 f_1 + \lambda_2 f_2|^p \, dx \le \frac{\lambda_1}{p} \int_a^b |f_1|^p \, dx + \frac{\lambda_2}{p} \int_a^b |f_2|^p \, dx$$
$$+ \left(1 - \frac{1}{p}\right) \int_a^b |\lambda_1 f_1 + \lambda_2 f_2|^p \, dx = 1 \, ;$$

hence, equality must hold in (I.14), (I.15), and (I.16). But the equality in (I.15) and (I.16) means that $|f_1| = |f_2|$ and equality in (I.14) that $f_1 f_2 \ge 0$; it follows that $f_1 = f_2$, contrary to our assumption. The proof is complete. ∎

In the case $p = 1$, the norm is no longer strictly convex. For example, let $f_1(x) = \frac{3}{2}x^2, f_2(x) = \frac{3}{4}(1 - x^2)$; then

$$\int_{-1}^1 |f_1(x)| \, dx = \int_{-1}^1 |f_2(x)| \, dx = 1,$$

and

$$\int_{-1}^1 \left| \frac{f_1(x) + f_2(x)}{2} \right| dx = \frac{3}{8} \int_{-1}^1 (1 + x^2) \, dx = 1.$$

The norm (I.12) is also strictly convex for $p > 1$. The proof proceeds exactly as in the case of the integral p-norm except that integration is replaced by summation over the points $x_1, \ldots, x_m$. When $p = 1$, the norm is not strictly convex and the uniqueness problem remains open.

The uniform norm (Examples I.1, I.2) is not strictly convex. For example, let $f_1(x) = x$, $f_2(x) = x^2$; then

$$\max_{0 \le x \le 1} |f_1(x)| = \max_{0 \le x \le 1} |f_2(x)| = 1$$

and

$$\max_{0 \le x \le 1} \left| \frac{f_1(x) + f_2(x)}{2} \right| = \frac{1}{2} \max_{0 \le x \le 1} |x + x^2| = 1.$$

The same is true for the discrete case (Example I.2). Thus, Theorem I.3 is uninformative in these cases, and the uniqueness question will have to receive special consideration.

Exercises

I.1   Show that $\| \cdot \|$ is a continuous function at each point v_0 of V in the sense that, given $\varepsilon > 0$, there exists $\delta > 0$ such that $|\, \|v\| - \|v_0\| \,| < \varepsilon$ whenever $\|v - v_0\| < \delta$. (In fact, $\delta = \varepsilon$.)

I.2   Show that

$$\int_a^b |f(x)g(x)| \, dx \le \left[\int_a^b |f(x)|^p \, dx \right]^{1/p} \cdot \left[\int_a^b |g(x)|^q \, dx \right]^{1/q},$$

where $p > 1$ and

$$(1/p) + (1/q) = 1.$$

Similarly, show that

$$\sum_{i=1}^n |a_i b_i| \le \left[\sum_{i=1}^n |a_i|^p \right]^{1/p} \cdot \left[\sum_{i=1}^n |b_i|^q \right]^{1/q}.$$

These inequalities are called Hölder's inequalities. The case $p = q = 2$ is usually called Schwarz's inequality.

[*Hint:* Use Lemma I.1.]

I.3   Show that if $p \ge 1$

$$\left[\int_a^b |f(x) + g(x)|^p \, dx \right]^{1/p} \le \left[\int_a^b |f(x)|^p \, dx \right]^{1/p} + \left[\int_a^b |g(x)|^p \, dx \right]^{1/p},$$

and

$$\left[\sum_{i=1}^n |a_i + b_i|^p \right]^{1/p} \le \left[\sum_{i=1}^n |a_i|^p \right]^{1/p} + \left[\sum_{i=1}^n |b_i|^p \right]^{1/p}.$$

These inequalities are called Minkowski's inequalities.

I.4 Show that (I.11) is, indeed, a norm in $C[a, b]$.

I.5 Show that (I.12) is not a strictly convex norm for $p = 1$.

I.6 Show that

$$\|f\| = \max_{i=1,\ldots,m} |f_i|$$

is not a strictly convex norm.

I.7 Prove: If V is a normed linear space, W a finite-dimensional subspace of V, and U a *closed* subset of W, then, given $v \in V$, there exists $u^* \in U$ such that $\|v - u^*\| \le \|v - u\|$ for all $u \in U$.

I.8 A trigonometric polynomial of degree, at most, n, is an expression of the form

$$t_n(\theta) = \sum_{k=0}^{n} (\alpha_k \cos k\theta + \beta_k \sin k\theta). \tag{I.17}$$

Show that, if $f(\theta)$ is continuous for $0 \le \theta \le 2\pi$, there exists a trigonometric polynomial of degree at most n, $t_n^*(\theta)$, such that

$$\max_{0 \le \theta \le 2\pi} |f(\theta) - t_n^*(\theta)| \le \max_{0 \le \theta \le 2\pi} |f(\theta) - t_n(\theta)|$$

for all t_n of the form (I.17).

I.9 The norm defined in (I.11) can be generalized by using a weight function. Namely, if $w(x)$ is a continuous nonnegative function on $[a, b]$, take

$$\|f\| = \left[\int_a^b |f(x)|^p w(x) \, dx \right]^{1/p}.$$

Show that this norm has all the properties we proved (I.11) to have.

I.10 Show that, if W is closed, so is W^* in Theorem I.2.

I.11 Show that in Theorem I.2 W^* is a bounded set. Indeed, if $w \in W^*$, then $\|w\| \le 2\|v\|$.

CHAPTER 1

UNIFORM APPROXIMATION

This chapter is devoted to the study of best approximations in the uniform norm. We first discuss how well continuous functions can be approximated by polynomials. Then we investigate the properties that characterize a best approximating polynomial. The scene next shifts to approximation on finite point sets and its relationship to approximation on an interval. This leads us into a discussion of computational methods for obtaining best uniform approximations numerically.

1.1 Uniform Approximation by Polynomials

1.1.1 The Weierstrass Theorem and Bernstein Polynomial Approximation

If P_n denotes the space of polynomials of degree at most n, then Theorem I.1 assures us that, given $f \in C[a, b]$, there exists a polynomial $p_n^* \in P_n$ such that

$$\|f - p_n^*\| \leq \|f - p\|, \qquad \text{all} \qquad p \in P_n,$$

where $\|\cdot\|$ is the uniform norm over the interval $[a, b]$, that is,

$$\|g\| = \max_{a \leq x \leq b} |g(x)|$$

for any $g \in C[a, b]$.

Let us put

$$E_n(f; [a, b]) = E_n(f) = \|f - p_n^*\|.$$

The first question we consider is: What is the behavior of $E_n(f)$ as $n \to \infty$? We shall show that $E_n(f) \to 0$ as $n \to \infty$ for each function $f(x)$ continuous on $[a, b]$. That is, a continuous function on a finite interval can be approximated uniformly within any preassigned error by polynomials. This result is the famous Weierstrass approximation theorem, which we state as follows.

THEOREM 1.1.[1]† *Given* $f(x)$ *continuous on* $[a, b]$ *and* $\varepsilon > 0$, *there exists a polynomial,* $p(x)$, *such that*

$$\|f(x) - p(x)\| < \varepsilon.$$

† Exponent numbers in parentheses refer to notes at the end of the chapters.

Proof. Following S. Bernstein, we shall construct the required polynomial explicitly. Let $x = (b - a)t + a$; then as x varies from a to b, t varies from 0 to 1. Put $f(x) = f((b - a)t + a) = g(t)$; then $g(t)$ is continuous on $[0, 1]$. Given any function, $h(t)$, which is bounded on $[0, 1]$, we define its *Bernstein Polynomial of degree m*, $m = 1, 2, \ldots, B_m(h; t)$, by

$$B_m(h; t) = \sum_{k=0}^{m} h\left(\frac{k}{m}\right)\binom{m}{k} t^k (1 - t)^{m-k}. \tag{1.1.1}$$

We note that $B_m(h; t) \in P_m$ and that

$$B_m(\alpha h; t) = \alpha B_m(h; t), \tag{1.1.2}$$

$$B_m(h_1 + h_2; t) = B_m(h_1; t) + B_m(h_2; t). \tag{1.1.3}$$

[Equations (1.1.2) and (1.1.3) imply that if we denote by B_m the *operator* which sends $h(t)$ into $B_m(h; t)$, then B_m is a *linear* operator.] Moreover, if $h_1(t) \le h_2(t)$ for all $t \in [0, 1]$, then for all $t \in [0, 1]$

$$B_m(h_1; t) \le B_m(h_2; t). \tag{1.1.4}$$

To establish (1.1.4), it is enough, in view of (1.1.2) and (1.1.3), to show that if $h(t) \ge 0$, then $B_m(h; t) \ge 0$, for (1.1.4) then follows, upon taking $h = h_2 - h_1$. But if $h(t) \ge 0$ for all $t \in [0, 1]$, then surely $h(k/m) \ge 0$, $k = 0, \ldots, m$, and $B_m(h; t)$ is a sum of nonnegative terms for $t \in [0, 1]$.

We wish to show that, given $\varepsilon > 0$ and $h(t) \in C[0, 1]$, there exists an integer m_0 such that

$$\|h - B_{m_0}(h)\| < \varepsilon. \tag{1.1.5}$$

We first consider some special cases. Suppose $h = 1$; then

$$B_m(1; t) = \sum_{k=0}^{m} \binom{m}{k} t^k (1 - t)^{m-k} = (t + (1 - t))^m = 1. \tag{1.1.6}$$

Next suppose $h(t)$ is t; then

$$B_m(t; t) = \sum_{k=0}^{m} \frac{k}{m} \binom{m}{k} t^k (1 - t)^{m-k} = \sum_{k=1}^{m} \frac{k}{m} \binom{m}{k} t^k (1 - t)^{m-k}.$$

But

$$\frac{k}{m}\binom{m}{k} = \binom{m-1}{k-1};$$

hence, putting $j = k - 1$, we obtain

$$B_m(t; t) = t \sum_{j=0}^{m-1} \binom{m-1}{j} t^j (1 - t)^{(m-1)-j} = t. \tag{1.1.7}$$

Finally, suppose $h(t)$ is t^2:

$$B_m\left(t\left(t - \frac{1}{m}\right); t\right) = \sum_{k=0}^{m} \frac{k}{m} \frac{k-1}{m} \binom{m}{k} t^k(1-t)^{m-k}$$

$$= \sum_{k=2}^{m} \frac{k}{m} \frac{k-1}{m} \binom{m}{k} t^k(1-t)^{m-k}.$$

But having

$$\frac{k}{m} \frac{k-1}{m} \binom{m}{k} = \left(1 - \frac{1}{m}\right)\binom{m-2}{k-2}$$

and putting $j = k - 2$, we obtain on the one hand

$$B_m\left(t\left(t - \frac{1}{m}\right); t\right) = \left(1 - \frac{1}{m}\right) t^2 \sum_{j=0}^{m-2} \binom{m-2}{j} t^j(1-t)^{m-2-j} = \left(1 - \frac{1}{m}\right)t^2$$

while on the other

$$B_m\left(t\left(t - \frac{1}{m}\right); t\right) = B_m(t^2; t) - \frac{1}{m} B_m(t; t) = B_m(t^2; t) - \frac{t}{m}.$$

Thus

$$B_m(t^2; t) = \frac{1}{m} t + \left(1 - \frac{1}{m}\right) t^2 = t^2 + \frac{1}{m} t(1-t). \tag{1.1.8}$$

These three instances are encouraging and, more important, they help us to prove our result for any continuous $h(t)$.

Suppose $\|h\| = M$ and $s, t \in [0, 1]$; then

$$-2M \le h(t) - h(s) \le 2M. \tag{1.1.9}$$

Also $h(t)$ is uniformly continuous on $[0, 1]$; hence, given $\varepsilon_1 > 0$, there exists $\delta(\varepsilon_1) > 0$ such that

$$-\varepsilon_1 < h(t) - h(s) < \varepsilon_1 \tag{1.1.10}$$

when

$$|t - s| < \delta.$$

(1.1.9) and (1.1.10) imply

$$-\varepsilon_1 - \frac{2M}{\delta^2}(t-s)^2 \le h(t) - h(s) \le \varepsilon_1 + \frac{2M}{\delta^2}(t-s)^2 \tag{1.1.11}$$

for $s, t \in [0, 1]$. For, if $|t - s| < \delta$, (1.1.10) implies (1.1.11) while, if $|t - s| \ge \delta$, then $(t - s)^2/\delta^2 \ge 1$ and (1.1.9) implies (1.1.11). If we fix s for the moment, take the Bernstein polynomials of degree m of the functions of t occurring in (1.1.11) and evaluate them at $t = s$, then we obtain, in view of (1.1.2), (1.1.3), (1.1.4), and (1.1.6),

$$-\varepsilon_1 - \frac{2M}{\delta^2} B_m((t-s)^2; s) \le B_m(h; s) - h(s)$$

$$\le \varepsilon_1 + \frac{2M}{\delta^2} B_m((t-s)^2; s). \tag{1.1.12}$$

But $(t - s)^2 = t^2 - 2st + s^2$; hence

$$B_m((t - s)^2; s) = B_m(t^2; s) - 2sB_m(t; s) + s^2B_m(1; s) = \frac{s(1 - s)}{m}$$

in view of (1.1.6), (1.1.7), and (1.1.8).

For $0 \le s \le 1$,

$$0 \le s(1 - s) \le \tfrac{1}{4};$$

hence, (1.1.12) implies that for $0 \le s \le 1$

$$|h(s) - B_m(h; s)| \le \varepsilon_1 + \frac{M}{2\delta^2 m}. \tag{1.1.13}$$

Now, if we choose $\varepsilon_1 = \varepsilon/2$, then (1.1.5) follows for any m_0 satisfying

$$m_0 > \frac{M}{\delta^2 \varepsilon}.$$

In particular, then,

$$|g(t) - B_{m_0}(g; t)| < \varepsilon, \qquad 0 \le t \le 1,$$

and the theorem is proved by taking $p(x) = B_{m_0}(g; (x - a)/(b - a))$.

The Bernstein polynomials provide us with explicit approximations to a given continuous function. We have just seen that the sequence $\{B_n(h; t)\}$ converges uniformly to the continuous $h(t)$ on $[0, 1]$. The next question to consider is: How good an approximation can be obtained out of P_n? We first obtain a bound for the error

$$\max_{0 \le t \le 1} |h(t) - B_n(h; t)|.$$

To this end, we need some more detailed information about continuous functions. Let $f(x)$ be defined on $[a, b]$, the *modulus of continuity* of $f(x)$ on $[a, b]$, $\omega(\delta)$, is defined for $\delta > 0$ by

$$\omega(\delta) = \sup_{\substack{x_1, x_2 \in [a,b], \\ |x_1 - x_2| \le \delta}} |f(x_1) - f(x_2)|.$$

Note that the modulus of continuity depends on δ, the function f, and the interval $[a, b]$, so that $\omega(\delta)$ is shorthand for $\omega(f; [a, b]; \delta)$. We need some properties of the modulus of continuity.

LEMMA 1.1. *If* $0 < \delta_1 \le \delta_2$, *then* $\omega(\delta_1) \le \omega(\delta_2)$.

LEMMA 1.2. *$f(x)$ is uniformly continuous on* $[a, b]$ *if and only if*

$$\lim_{\delta \to 0} \omega(\delta) = 0.$$

We leave the proofs of these results to the reader.

LEMMA 1.3.   *If* $\lambda > 0$, *then*

$$\omega(\lambda\delta) \leq (1 + \lambda)\omega(\delta).$$

Proof.   Let n be an integer such that $n \leq \lambda < n + 1$; then $\omega(\lambda\delta) \leq \omega((n + 1)\delta)$. Suppose $|x_1 - x_2| \leq (n + 1)\delta$ and $x_2 > x_1$, say. We divide the interval $[x_1, x_2]$ into $n + 1$ equal parts each of length $(x_2 - x_1)/(n + 1)$ by means of the points

$$z_j = x_1 + j(x_2 - x_1)/(n + 1), \qquad j = 0, 1, \ldots, n + 1.$$

Then

$$|f(x_2) - f(x_1)| = \left|\sum_{j=0}^{n} [f(z_{j+1}) - f(z_j)]\right| \leq \sum_{j=0}^{n} |f(z_{j+1}) - f(z_j)| \leq (n + 1)\omega(\delta).$$

Thus, $\omega((n + 1)\delta) \leq (n + 1)\omega(\delta)$. But $n + 1 \leq \lambda + 1$, and the lemma is proved. ∎

THEOREM 1.2.   *If* $h(t)$ *is bounded for* $0 \leq t \leq 1$, *then*

$$\|h - B_n(h)\| \leq \frac{3}{2}\,\omega\left(\frac{1}{\sqrt{n}}\right). \tag{1.1.14}$$

Proof.   In view of (1.1.6) and (1.1.2)

$$|h(t) - B_n(h; t)| = \left|\sum_{k=0}^{n} \left[h(t) - h\left(\frac{k}{n}\right)\right] \binom{n}{k} t^k(1 - t)^{n-k}\right|$$

$$\leq \sum_{k=0}^{n} \left|h(t) - h\left(\frac{k}{n}\right)\right| \binom{n}{k} t^k(1 - t)^{n-k}$$

$$\leq \sum_{k=0}^{n} \omega\left(\left|t - \frac{k}{n}\right|\right) \binom{n}{k} t^k(1 - t)^{n-k}.$$

Now, applying Lemma 1.3,

$$\omega\left(\left|t - \frac{k}{n}\right|\right) = \omega\left(n^{1/2}\left|t - \frac{k}{n}\right| n^{-1/2}\right) \leq \left(1 + n^{1/2}\left|t \cdot \frac{k}{n}\right|\right) \omega(n^{-1/2}),$$

so that

$$|h(t) - B_n(h; t)| \leq \sum_{k=0}^{n} \left(1 + n^{1/2}\left|t - \frac{k}{n}\right|\right) \omega(n^{-1/2}) \binom{n}{k} t^k(1 - t)^{n-k}$$

$$\leq \omega(n^{-1/2}) \left[1 + n^{1/2} \sum_{k=0}^{n} \left|t - \frac{k}{n}\right| \binom{n}{k} t^k(1 - t)^{n-k}\right].$$

But Schwarz's inequality (see Exercise I.2) implies that

$$\sum_{k=0}^{n} \left| t - \frac{k}{n} \right| \binom{n}{k} t^k (1-t)^{n-k}$$

$$= \sum_{k=0}^{n} \left(\left| t - \frac{k}{n} \right| \sqrt{\binom{n}{k} t^k (1-t)^{n-k}} \right) \left(\sqrt{\binom{n}{k} t^k (1-t)^{n-k}} \right)$$

$$\leq \left[\sum_{k=0}^{n} \left(t - \frac{k}{n} \right)^2 \binom{n}{k} t^k (1-t)^{n-k} \right]^{1/2} \left[\sum_{k=0}^{n} \binom{n}{k} t^k (1-t)^{n-k} \right]^{1/2}$$

$$\leq \left[\sum_{k=0}^{n} \left(t - \frac{k}{n} \right)^2 \binom{n}{k} t^k (1-t)^{n-k} \right]^{1/2}$$

and

$$\sum_{k=0}^{n} \left(t - \frac{k}{n} \right)^2 \binom{n}{k} t^k (1-t)^{n-k} = \frac{t(1-t)}{n} \leq \frac{1}{4n},$$

as we saw in the course of the proof of Theorem 1.1. Thus,

$$|h(t) - B_n(h; t)| \leq \omega(n^{-1/2}) \left[1 + n^{1/2} \cdot \frac{1}{2n^{1/2}} \right],$$

and the theorem is proved. ∎

Remarks. 1. If $h(t)$ is continuous on $[0, 1]$, in view of Lemma 1.2, we see that $\omega(n^{-1/2}) \to 0$ as $n \to \infty$, and we have another proof of Theorem 1.1.

2. If a function satisfies

$$|f(x_1) - f(x_2)| \leq K|x_1 - x_2|^\alpha$$

for $x_1, x_2 \in [a, b]$, and $0 < \alpha$, then $f(x)$ is said to satisfy a Lipschitz condition of order α with constant K on $[a, b]$, and the set of such functions is denoted by $\text{lip}_K \alpha$, or $\text{lip} \alpha$ if the constant K is unimportant. It is easy to see that $f(x) \in \text{lip}_K \alpha$ if and only if $\omega(\delta) \leq K\delta^\alpha$ and, hence, if $h(t) \in \text{lip}_K \alpha$ on $[0, 1]$:

$$\|h - B_n(h)\| \leq \tfrac{3}{2} K n^{-\alpha/2}. \tag{1.1.15}$$

(Of course, if $\alpha > 1$ and $f \in \text{lip} \alpha$, then f' exists on $[a, b]$ and is identically zero. Thus, for $\alpha > 1$, $\text{lip} \alpha$ consists of constants, therefore, the Lipschitz condition has interest only for $0 < \alpha \leq 1$.)

3. Suppose that $h(t) = \left| t - \frac{1}{2} \right|$. It is easy to verify that $h(t) \in \text{lip}_1 1$, so that $\|h - B_n(h)\| \leq \frac{3}{2} n^{-1/2}$. But let us consider the point $t = \frac{1}{2}$. At $t = \frac{1}{2}$,

$$B_n(h; t) - \left| t - \frac{1}{2} \right| = \left(\frac{1}{2} \right)^n \sum_{k=0}^{n} \left| \frac{k}{n} - \frac{1}{2} \right| \binom{n}{k}.$$

Suppose that n is even; then

$$\sum_{k=0}^{n} \left| \frac{k}{n} - \frac{1}{2} \right| \binom{n}{k} = 2 \sum_{k=0}^{n/2} \left(\frac{1}{2} - \frac{k}{n} \right) \binom{n}{k} = \frac{1}{2} \binom{n}{n/2},$$

and so

$$\left| B_n \left(h; \frac{1}{2} \right) - h \left(\frac{1}{2} \right) \right| = \frac{\binom{n}{n/2}}{2^{n+1}} > \frac{1}{2} n^{-1/2}.$$

The final inequality is a consequence of a sharp form of Stirling's formula (cf. Courant, vol. I, p. 361),

$$\sqrt{2\pi k}\, k^k e^{-k} < k! < \sqrt{2\pi k}\, k^k e^{-k} \left(1 + \frac{1}{4k} \right).$$

Thus,

$$\| h - B_n(h) \| > \tfrac{1}{2} n^{-1/2}.$$

We, therefore, cannot expect to do better than (1.1.14), at least as far as the exponent of n goes, using Bernstein polynomials.

1.1.2 Jackson's Theorems

Our next result, due to D. Jackson, tells us that we can improve on (1.1.14) and, indeed, gives the correct asymptotic behavior of $E_n(f)$ as $n \to \infty$. This is a substantial result and will require some preparatory work.

Let $g(\theta)$ be a function continuous for $-\pi \le \theta \le \pi$ and periodic with period 2π. We associate with $g(\theta)$ a trigonometric polynomial of degree n,

$$s_n(g; \theta) = \frac{a_0}{2} + \sum_{k=1}^{n} (a_k \cos k\theta + b_k \sin k\theta), \qquad (1.1.16)$$

where the coefficients are determined by

$$a_k = \frac{1}{\pi} \int_{-\pi}^{\pi} g(\phi) \cos k\phi \, d\phi, \qquad k = 0, 1, \ldots, n,$$

$$b_k = \frac{1}{\pi} \int_{-\pi}^{\pi} g(\phi) \sin k\phi \, d\phi, \qquad k = 1, \ldots, n. \tag{1.1.17}$$

The trigonometric polynomial $s_n(g; \theta)$ is nothing more than the partial sum, through terms of order n, of the Fourier series of $g(\theta)$. We shall require trigonometric polynomials that are somewhat more general than $s_n(g; \theta)$. Namely, we define

$$q_n(g; \theta) = \frac{a_0}{2} + \sum_{k=1}^{n} \rho_{k,n}(a_k \cos k\theta + b_k \sin k\theta), \qquad (1.1.18)$$

where $a_0, a_1, \ldots, a_n; b_1, \ldots, b_n$ are the Fourier coefficients of $g(\theta)$ defined in
(1.1.17) and $\rho_{1,n}, \ldots, \rho_{n,n}$, $n = 1, 2, \ldots$ are any given real numbers.

LEMMA 1.4. *If $g(\theta)$ is continuous on $-\pi \leq \theta \leq \pi$ and has period 2π, then*

$$q_n(g; \theta) = \frac{1}{\pi} \int_{-\pi}^{\pi} g(\phi + \theta) u_n(\phi) \, d\phi, \tag{1.1.19}$$

where

$$u_n(\phi) = \frac{1}{2} + \sum_{k=1}^{n} \rho_{k,n} \cos k\phi. \tag{1.1.20}$$

Proof. If we substitute (1.1.17) into (1.1.18) and recall the identity
$\cos A \cos B + \sin A \sin B \equiv \cos (A - B)$, we obtain

$$q_n(g; \theta) = \frac{1}{\pi} \int_{-\pi}^{\pi} g(\phi) u_n(\phi - \theta) \, d\phi = \frac{1}{\pi} \int_{-\pi-\theta}^{\pi-\theta} g(\tau + \theta) u_n(\tau) \, d\tau.$$

But both g and u_n have period 2π; hence

$$\frac{1}{\pi} \int_{-\pi-\theta}^{-\pi} g(\tau + \theta) u_n(\tau) \, d\tau = \frac{1}{\pi} \int_{\pi-\theta}^{\pi} g(\tau + \theta) u_n(\tau) \, d\tau.$$

Thus,

$$q_n(g; \theta) = \frac{1}{\pi} \int_{-\pi}^{\pi} g(\tau + \theta) u_n(\tau) \, d\tau,$$

which was to be proved. ∎

LEMMA 1.5

$$|\theta| \leq \frac{\pi}{2} |\sin \theta| \qquad \text{for} \qquad 0 \leq |\theta| \leq \frac{\pi}{2}.$$

Proof. The second derivative of $-\sin \theta$ is positive for $0 < \theta \leq \pi/2$; hence
$-\sin \theta$ is a convex function and the point $(\theta, -(2/\pi)\theta)$ of the chord joining
$(0, 0)$ and $(\pi/2, -1)$ cannot be below the point $(\theta, -\sin \theta)$. Thus, $-(2/\pi)\theta \geq$
$-\sin \theta$ for $0 \leq \theta \leq \pi/2$, and the lemma follows. ∎

LEMMA 1.6

$$\sin \theta \leq \theta, \qquad \theta \geq 0.$$

Proof. Let $k(\theta) = \theta - \sin \theta$. Then by the mean-value theorem there
exists ξ, $0 \leq \xi \leq \theta$ such that $k(\theta) - k(0) = \theta k'(\xi)$ or $\theta - \sin \theta =$
$\theta(1 - \cos \xi) \geq 0$. ∎

LEMMA 1.7. *Suppose $\rho_{1,n}, \ldots, \rho_{n,n}$ to be chosen in such a way that*

$$u_n(\phi) \geq 0, \qquad -\pi \leq \phi \leq \pi; \tag{1.1.21}$$

then, for $-\pi \le \theta \le \pi$,

$$|g(\theta) - q_n(g; \theta)| \le \omega\left(\frac{1}{n}\right)\left[1 + \frac{n\pi}{\sqrt{2}}(1 - \rho_{1,n})^{1/2}\right], \qquad (1.1.22)$$

where $\omega(1/n)$ *is short for* $\omega(g; [-\pi, \pi]; 1/n)$.

Proof. In view of (1.1.19) and the fact that $(1/\pi)\int_{-\pi}^{\pi} u_n(\phi)\, d\phi = 1$,

$$|g(\theta) - q_n(g; \theta)| = \left|\frac{1}{\pi}\int_{-\pi}^{\pi}[g(\theta) - g(\phi + \theta)]u_n(\phi)\, d\phi\right| \le \frac{1}{\pi}\int_{-\pi}^{\pi}\omega(|\phi|)u_n(\phi)\, d\phi,$$

where $\omega(|\phi|)$ is short for $\omega(g; [-\pi, \pi], |\phi|)$. Note that for a given θ and $-\pi \le \phi \le \pi$,

$$|g(\theta) - g(\theta + \phi)| \le \omega(g; [\theta - \pi, \theta + \pi]; |\phi|) = \omega(g; [-\pi, \pi]; |\phi|)$$

(cf. Exercise 1.5). We now repeat the device used in the proof of Theorem 1.2.

$$\omega(|\phi|) = \omega\left(n|\phi| \cdot \frac{1}{n}\right) \le (1 + n|\phi|)\omega\left(\frac{1}{n}\right)$$

and, hence,

$$|g(\theta) - q_n(g; \theta)| \le \omega\left(\frac{1}{n}\right)\left[1 + \frac{n}{\pi}\int_{-\pi}^{\pi}|\phi|u_n(\phi)\, d\phi\right]. \qquad (1.1.23)$$

Now, by Schwarz's inequality for integrals,

$$\frac{1}{\pi}\int_{-\pi}^{\pi}|\phi|u_n(\phi)\, d\phi = \frac{1}{\pi}\int_{-\pi}^{\pi}|\phi|[u_n(\phi)]^{1/2} \cdot ([u_n(\phi)]^{1/2})\, d\phi$$

$$\le \left[\frac{1}{\pi}\int_{-\pi}^{\pi}\phi^2 u_n(\phi)\, d\phi\right]^{1/2} \cdot \left[\frac{1}{\pi}\int_{-\pi}^{\pi}u_n(\phi)\, d\phi\right]^{1/2}$$

$$\le \left[\frac{1}{\pi}\int_{-\pi}^{\pi}\phi^2 u_n(\phi)\, d\phi\right]^{1/2}. \qquad (1.1.24)$$

According to Lemma 1.5, if $-\pi \le \phi \le \pi$,

$$\left(\frac{\phi}{2}\right)^2 \le \frac{\pi^2}{4}\sin^2\frac{\phi}{2} = \frac{\pi^2}{4}\frac{(1 - \cos\phi)}{2}$$

or

$$\phi^2 \le \pi^2\frac{(1 - \cos\phi)}{2}.$$

Thus, (1.1.24) implies

$$\frac{1}{\pi} \int_{-\pi}^{\pi} |\phi| u_n(\phi) \, d\phi \leq \left[\frac{\pi}{2} \int_{-\pi}^{\pi} (1 - \cos \phi) u_n(\phi) \, d\phi \right]^{1/2} = \frac{\pi}{\sqrt{2}} (1 - \rho_{1,n})^{1/2}$$

(1.1.25)

in view of the identity $\cos \phi \cos k\phi \equiv \frac{1}{2}[\cos (k - 1)\phi + \cos (k + 1)\phi]$, $k = 1, 2, \ldots,$. The lemma is now proved by substituting (1.1.25) into (1.1.23). ∎

Our next task is to find a particular set $\rho_{1,n}, \ldots, \rho_{n,n}$ such that the resulting $u_n(\phi)$ is nonnegative while $\rho_{1,n}$ is close to 1. An easy way to produce non-negative cosine polynomials is to notice that for any real numbers $c_0, c_1, \ldots, c_n$, on the one hand,

$$\left(\sum_{k=0}^{n} c_k e^{ik\theta} \right) \left(\sum_{k=0}^{n} c_k e^{-ik\theta} \right) = \left| \sum_{k=0}^{n} c_k e^{ik\theta} \right|^2 \geq 0$$

(we are relying on the fact that for any complex number z, $z\bar{z} = |z|^2$; $\bar{z}$ is the complex conjugate of z, for example, $e^{-ik\theta} = \overline{e^{ik\theta}}$). On the other hand,

$$\left(\sum_{k=0}^{n} c_k e^{ik\theta} \right) \left(\sum_{k=0}^{n} c^k e^{-ik\theta} \right) = \sum_{k=0}^{n} c_k^2 + \left(2 \sum_{k=0}^{n-1} c_k c_{k+1} \right) \cos \theta$$

$$+ \left(2 \sum_{k=0}^{n-2} c_k c_{k+2} \right) \cos 2\theta + \cdots$$

$$+ \left(2 \sum_{k=0}^{n-p} c_k c_{k+p} \right) \cos p\theta + \cdots + 2c_0 c_n \cos n\theta.$$

To produce a $u_n(\phi)$, we must have $c_0^2 + c_1^2 + \cdots + c_n^2 = \frac{1}{2}$. Let us choose

$$c_k = c \sin \frac{k + 1}{n + 2} \pi, \qquad k = 0, \ldots, n,$$

where c satisfies

$$c^2 = \frac{1}{2 \sum_{k=0}^{n} \sin^2 [(k + 1)/(n + 2)] \pi}.$$

Then

$$c_0^2 + c_1^2 + \cdots + c_n^2 = \frac{1}{2}$$

and we have obtained a nonnegative $u_n(\phi)$. What is

$$\rho_{1,n} = 2c^2 \sum_{k=0}^{n-1} \left(\sin \frac{k + 1}{n + 2} \pi \right) \left(\sin \frac{k + 2}{n + 2} \pi \right)?$$

Note that

$$\sum_{k=0}^{n-1} \left(\sin \frac{k+1}{n+2} \pi \right) \left(\sin \frac{k+2}{n+2} \pi \right)$$

$$= \sum_{k=0}^{n} \left(\sin \frac{k+1}{n+2} \pi \right) \left(\sin \frac{k+2}{n+2} \pi \right)$$

$$= \sum_{k=0}^{n} \left(\sin \frac{k\pi}{n+2} \right) \left(\sin \frac{k+1}{n+2} \pi \right)$$

$$= \frac{1}{2} \sum_{k=0}^{n} \left(\sin \frac{k+2}{n+2} \pi + \sin \frac{k\pi}{n+2} \right) \sin \frac{k+1}{n+2} \pi$$

$$= \sum_{k=0}^{n} \left(\cos \frac{\pi}{n+2} \right) \left(\sin \frac{k+1}{n+2} \pi \right) \left(\sin \frac{k+1}{n+2} \pi \right)$$

$$= \cos \frac{\pi}{n+2} \sum_{k=0}^{n} \sin^2 \frac{k+1}{n+2} \pi = \frac{1}{2c^2} \cos \frac{\pi}{n+2},$$

where we have used the identity

$$\sin A + \sin B \equiv 2(\cos (A - B)/2)(\sin (A + B)/2).$$

Hence,

$$\rho_{1,n} = \cos \frac{\pi}{n+2},$$

but since

$$(1 - \rho_{1,n})^{1/2} = \left(1 - \cos \frac{\pi}{n+2} \right)^{1/2} = \sqrt{2} \sin \frac{\pi}{2n+4} \leq \sqrt{2} \frac{\pi}{2n+4}$$

in view of Lemma 1.6,

$$1 + \frac{n\pi}{\sqrt{2}} (1 - \rho_{1,n})^{1/2} \leq 1 + \frac{n\pi^2}{2n+4} \leq 1 + \frac{\pi^2}{2} \leq 6.$$

Thus, by making this particular choice of $u_n(\phi)$, we obtain from Lemma 1.7

$$|g(\theta) - q_n(g; \theta)| \leq 6\omega \left(g; [-\pi, \pi]; \frac{1}{n} \right), \qquad -\pi \leq \theta \leq \pi. \qquad (1.1.26)$$

According to Exercise I.8, $g(\theta)$ has a best uniform approximation among trigonometric polynomials of degree at most n, call it $q_n^*(\theta)$. Therefore, we have proved

THEOREM 1.3. *If $g(\theta)$ is continuous on $[-\pi, \pi]$ and has period 2π, then*

$$\|g - q_n^*\| \leq 6\omega \left(\frac{1}{n} \right). \qquad (1.1.27)$$

Suppose that $f(x) \in C[-1, 1]$. Then $g(\theta) = f(\cos \theta)$ is continuous on $0 \leq \theta \leq \pi$, and we define it on $-\pi \leq \theta < 0$ by $g(\theta) = g(-\theta)$ to obtain an *even* function, continuous on $[-\pi, \pi]$ having period 2π. Since $g(\theta)$ is *even*, it has a best uniform approximation by trigonometric polynomials of degree at most n which is also even (cf. Exercise 1.1), call it q_n^*. An *even* trigonometric polynomial of degree n has the form

$$q_n^*(\theta) = \frac{a_0}{2} + a_1 \cos \theta + \cdots + a_n \cos n\theta.$$

That is, all the sine terms have zero coefficients. It is easy to verify (cf. Exercise 1.6) that $\cos k\theta$ is a polynomial in $\cos \theta$ of degree at most k; hence,

$$q_n^*(\theta) = d_0 + d_1 \cos \theta + d_2(\cos \theta)^2 + \cdots + d_n(\cos \theta)^n,$$

and, if we put

$$p_n^*(x) = d_0 + d_1 x + \cdots + d_n x^n,$$

then according to (1.1.26) and Exercise 1.4

$$\max_{-1 \leq x \leq 1} |f(x) - p_n^*(x)| \leq 6\omega \left(g; [-\pi, \pi]; \frac{1}{n} \right) \leq 6\omega \left(f; [-1, 1]; \frac{1}{n} \right).$$

Finally, then, we have established

THEOREM 1.4[(2)] (JACKSON'S THEOREM). *If* $f \in C[-1, 1]$, *then*

$$E_n(f; [-1, 1]) \leq 6\omega \left(\frac{1}{n} \right). \tag{1.1.28}$$

This is our main result, and we turn at once to its immediate consequences.

COROLLARY 1.4.1. *If* $f \in C[a, b]$, *then*

$$E_n(f; [a, b]) \leq 6\omega \left(\frac{b - a}{2n} \right).$$

Proof. Apply Exercise 1.7 to the result of Theorem 1.4.

COROLLARY 1.4.2. *If* $f \in \text{lip}_K \alpha$ *on* $[-1, 1]$, *then*

$$E_n(f; [-1, 1]) \leq 6Kn^{-\alpha}.$$

COROLLARY 1.4.3. *If* $|f'(x)| \leq M$ *for* $-1 \leq x \leq 1$, *then*

$$E_n(f; [-1, 1]) \leq 6Mn^{-1}.$$

Proof. See Exercise 1.8.

COROLLARY 1.4.4. *If $f'(x) \in C[-1, 1]$, then*

$$E_n(f; [-1, 1]) \le 6E_{n-1}(f'; [-1, 1])n^{-1}.$$

Proof. Suppose that $\|f' - p_{n-1}^*\| = E_{n-1}(f'; [-1, 1])$. Put

$$p_n(x) = \int_0^x p_{n-1}^*(t)\, dt.$$

Then (see Exercise 1.9)

$$E_n(f; [-1, 1]) = E_n(f - p_n; [-1, 1]) \le 6E_{n-1}(f'; [-1, 1])n^{-1},$$

where we use Corollary 1.4.3 and the fact that $p_n' = p_{n-1}^*$.

THEOREM 1.5. *Let $f(x)$ have a kth derivative on $[-1, 1]$; then, if $n > k$,*

$$E_n(f; [-1, 1]) \le \frac{c}{n^k} \omega_k \left(\frac{1}{n - k} \right),$$

where ω_k is the modulus of continuity of $f^{(k)}$ and

$$c = 6^{k+1}e^k(1 + k)^{-1}.$$

Proof. By repeated application of Corollary 1.4.4 we obtain

$$E_{n-j}(f^{(j)}; [-1, 1]) \le 6E_{n-j-1}(f^{(j+1)}; [-1, 1])(n - j)^{-1},$$
$$j = 0, \ldots, k - 1.$$

Hence, multiplying these inequalities yields

$$E_n(f)E_{n-1}(f') \cdots E_{n-(k-1)}(f^{(k-1)})$$
$$\le 6^k[n(n - 1) \cdots (n - (k - 1))]^{-1}E_{n-1}(f')E_{n-2}(f'') \cdots E_{n-k}(f^{(k)}).$$
$$(1.1.29)$$

If $E_{n-j}(f^{(j)}) = 0$ for any j, then $E_n(f) = 0$ and the theorem is trivially true. Suppose that $E_{n-j}(f^{(j)}) \ne 0, j = 0, \ldots, k - 1$. Then (1.1.29) implies

$$E_n(f) \le 6^k[n(n - 1) \cdots (n - (k - 1))]^{-1}E_{n-k}(f^{(k)}).$$

But, in view of Theorem 1.4, $E_{n-k}(f^{(k)}) \le 6\omega_k((n - k)^{-1})$ hence,

$$E_n(f) \le 6^{k+1}[n(n - 1) \cdots (n - (k - 1))]^{-1}\omega_k((n - k)^{-1}).$$

Now, it is an elementary observation that

$$\log[n(n - 1) \cdots (n - (k - 1))] = \log n + \log(n - 1) + \cdots$$
$$+ \log(n - (k - 1)) \ge \int_{n-k}^n \log t\, dt,$$

from which we conclude that

$$[n(n - 1) \cdots (n - (k - 1))] \ge n^k e^{-k} \left(1 + \frac{k}{n - k} \right)^{n-k} \ge n^k e^{-k}(1 + k)$$

and, therefore,

$$[n(n - 1) \cdots (n - (k - 1))]^{-1} \leq n^{-k}e^k(1 + k)^{-1}. \tag{1.1.30}$$

The theorem now follows. ∎

There are results corresponding to Corollaries 1.4.2, 1.4.3, and Theorem 1.5 for any finite interval $[a, b]$ easily derivable from Corollary 1.4.1, and the reader interested in them is invited to provide them.

We have spent a good deal of time on Theorem 1.4 and its consequences because (1.1.28) is the best possible bound on $E_n(f)$, in the sense that, very roughly speaking, if a continuous f is not k times differentiable, then $E_n(f)$ cannot satisfy an inequality of the form of (1.1.28) for $n = 1, 2, \ldots$. There is, indeed, a precise inverse theory, due to S. Bernstein, as the previous sentence suggests. The reader may consult Cheney [1] and Timan [1] and the literature cited there for these details, as well as such related facts as what the smallest number is that can replace 6 in Theorem 1.4, what the smallest value of c is for which Theorem 1.5 remains true, and other similar refinements and embellishments.

To sum up, Theorem 1.5 gives us fairly precise information about how closely a function with known properties can be approximated by polynomials. It, therefore, provides us with a limitation on what we may expect from polynomial approximation, but is of no help in actually obtaining best approximations. To this latter end, we shall next study the characteristic properties of best approximations.

1.2 Characterization of Best Approximations

We embark now on our study of the properties of polynomials, p_n^*, of best approximation to a given continuous $f(x)$ on the interval $[a, b]$. Let

$$e(x) = f(x) - p_n^*(x) \, ;$$

then $\|e(x)\| = E_n(f; [a, b])$.

THEOREM 1.6.   *There exist (at least) two distinct points* $x_1, x_2 \in [a, b]$ *such that*

$$|e(x_1)| = |e(x_2)| = E_n(f; [a, b]) \tag{1.2.1}$$

and

$$e(x_1) = -e(x_2). \tag{1.2.2}$$

Proof.   The continuous curve $y = e(x)$ is constrained to lie between the lines $y = \pm E_n(f)$ for $a \leq x \leq b$ and touches at least one of these lines. We wish to prove that it must touch *both* of them. If it does not, then (as Figure 1.1 suggests) a better approximation to f than p_n^* exists. For assume that $e(x) > -E_n(f)$ throughout $[a, b]$; then

$$\min_{a \le x \le b} e(x) = m > -E_n(f) \quad \text{and} \quad c = \frac{E_n(f) + m}{2} > 0.$$

Since $q_n = p_n^* + c \in P_n, f(x) - q_n(x) = e(x) - c$, and

$$-(E_n(f) - c) = m - c \le e(x) - c \le E_n(f) - c,$$

we have

$$\|f - q_n\| = E_n(f) - c,$$

contradicting the definition of $E_n(f)$. Thus, there must exist a point of $[a, b]$, call it x_1, such that $e(x_1) = -E_n(f)$. A similar argument establishes the existence of $x_2 \in [a, b]$ such that $e(x_2) = E_n(f)$, and the theorem is proved. ∎

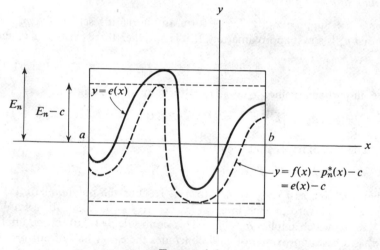

FIGURE 1.1

COROLLARY 1.6.1. *The best approximating constant to $f(x)$ is*

$$p_0^* = \tfrac{1}{2} \Big[\max_{a \le x \le b} f(x) + \min_{a \le x \le b} f(x) \Big] \tag{1.2.3}$$

and

$$E_0(f) = \tfrac{1}{2} \Big[\max_{a \le x \le b} f(x) - \min_{a \le x \le b} f(x) \Big].$$

Proof. If d is any constant other than p_0^*, $e(x) = f(x) - d$ cannot satisfy the conditions of Theorem 1.6. ∎

Theorem 1.6 is just a foreshadowing of the true state of affairs, however. As we shall show next, the curve $y = e(x)$ must touch the lines $y = \pm E_n(f)$ alternately at least $\underline{n + 2}$ times, and this property characterizes the best uniform approximation of a continuous function by a polynomial of degree at most n. A set of $k + 1$ distinct points $x_0, \ldots, x_k$, satisfying $a \leq x_0 < x_1 < \cdots < x_{k-1} < x_k \leq b$ is called an *alternating set* for the error function $f - p_n$ if

$$|f(x_j) - p_n(x_j)| = \|f - p_n\|, \qquad j = 0, \ldots, k \qquad (1.2.4)$$

and

$$[f(x_j) - p_n(x_j)] = -[f(x_{j+1}) - p_n(x_{j+1})], \qquad j = 0, \ldots, k - 1. \qquad (1.2.5)$$

THEOREM 1.7. *Suppose $f \in C[a, b]$; p_n^* is a best uniform approximation on $[a, b]$ to f out of P_n if and only if there exists an alternating set for $f - p_n^*$ consisting of $n + 2$ points.*

Proof. (i) Suppose $x_0, \ldots, x_{n+1}$ form an alternating set for $f - p_n^*$. We show that p_n^* is a best approximation. If it is not, then there exists $q_n \in P_n$ such that

$$\|f - q_n\| < \|f - p_n^*\|. \qquad (1.2.6)$$

Hence, in particular, since $x_0, \ldots, x_{n+1}$ form an alternating set,

$$|f(x_j) - q_n(x_j)| < \|f - p_n^*\| = |f(x_j) - p_n^*(x_j)|, \qquad j = 0, \ldots, n + 1. \qquad (1.2.7)$$

(1.2.7) and (1.2.5) imply that the difference

$$[f(x_j) - p_n^*(x_j)] - [f(x_j) - q_n(x_j)] = q_n(x_j) - p_n^*(x_j)$$

alternates in sign as j runs from 0 to $n + 1$. Thus the polynomial $q_n(x) - p_n^*(x) \in P_n$ has a zero in each interval (x_j, x_{j+1}), $j = 0, \ldots, n$, for a total of $n + 1$ zeros, which implies $q_n = p_n^*$. This contradicts (1.2.6), hence implies that p_n^* is a best approximation and concludes the easier half of our proof.

(ii) Suppose that p_n^* is a best approximation to f and $f \notin P_n$. (If $f \in P_n$, the whole question is trivial.) Let a largest alternating set for $f - p_n^*$ consist of the $k + 1$ points $x_0, \ldots, x_k$ satisfying $a \leq x_0 < x_1 < \cdots < x_{k-1} < x_k \leq b$. In view of Theorem 1.6, $k \geq 1$. We wish to prove that $k \geq n + 1$. Suppose, then, that $k \leq n$, and let us put

$$\|f - p_n^*\| = \rho \qquad (> 0).$$

Let $t_0, \ldots, t_s$ be points of $[a, b]$ chosen so that $a = t_0 < t_1 < \cdots < t_s = b$ and so that $e(x) = f(x) - p_n^*(x)$ satisfies

$$|e(\xi) - e(\eta)| < \tfrac{1}{2}\rho \qquad (1.2.8)$$

for $\xi, \eta \in [t_j, t_{j+1}]$, $j = 0, \ldots, s - 1$. If a subinterval $[t_j, t_{j+1}]$ contains a point, t, at which $e(t) = \rho$, we call it a $(+)$ subinterval, while if $e(t) = -\rho$ we call it a $(-)$ subinterval. Because of (1.2.8), $e(x) > 0$ throughout a $(+)$ subinterval and $e(x) < 0$ throughout a $(-)$ subinterval. Let us write down the subintervals, $[t_j, t_{j+1}]$, that are $(\pm)$ subintervals, in order, (from left to right) as $I_1, I_2, \ldots, I_N$, and suppose that I_1 is a $(+)$ subinterval. Next we divide $I_1, \ldots, I_N$ into subsets as follows:

$$\{I_1, I_2, \ldots, I_{k_1}\}: \qquad (+) \quad \text{subintervals,}$$
$$\{I_{k_1+1}, I_{k_1+2}, \ldots, I_{k_2}\}: \qquad (-) \quad \text{subintervals,}$$
$$\{I_{k_2+1}, I_{k_2+2}, \ldots, I_{k_3}\}: \qquad (+) \quad \text{subintervals,}$$
$$\vdots$$
$$\{I_{k_m+1}, I_{k_m+2}, \ldots, I_{k_{m+1}}\}: (-)^m \text{ subintervals.}$$

Each subset contains at least one element, and $2 \leq m + 1 \leq n + 1$. It is clear that I_{k_j} and I_{k_j+1} are disjoint for $j = 1, \ldots, m$. We can therefore choose points $z_1, \ldots, z_m$ with the property that $z_j > x$ for all $x \in I_{k_j}$ and $z_j < x$ for all $x \in I_{k_j+1}$. If we set

$$q(x) = (z_1 - x)(z_2 - x) \cdots (z_m - x),$$

then $q \in P_n$ and $q(x)$ has the same sign as $e(x)$ in each I_j, $j = 1, \ldots, N$.

Let R be the sum of all subintervals which are not $(\pm)$ subintervals; then

$$\max_{x \in R} |e(x)| = \rho' < \rho.$$

Suppose, further, that

$$\|q\| = M$$

and that $\lambda > 0$ is chosen so small that

$$\lambda M < \min (\rho - \rho', \rho/2).$$

We now claim that $p(x) = \lambda q(x) + p_n^*(x) \in P_n$ is a better approximation to $f(x)$ than $p_n^*(x)$.

If, on the one hand, $x \in R$, then

$$|f(x) - p(x)| = |e(x) - \lambda q(x)| \leq |e(x)| + \lambda |q(x)| \leq \rho' + \lambda M < \rho. \quad (1.2.9)$$

On the other hand, if $x \in I_j$, $j = 1, \ldots, N$ then $e(x)$ and $\lambda q(x)$ have the same sign (neither is zero), $|e(x)| > \rho/2 > |\lambda q(x)|$, and we have

$$|f(x) - p(x)| = |e(x) - \lambda q(x)| = |e(x)| - |\lambda q(x)| \leq \rho - |\lambda q(x)| < \rho.$$

Together with (1.2.9) this implies that

$$\|f - p\| < \rho = \|f - p_n^*\|.$$

This contradiction proves the theorem.[3] ∎

Note that the theorem does not say that the alternating set consisting of $n + 2$ points for $f - p_n^*$ is unique, nor does it imply that $f - p_n^*$ may not "alternate" on more than $n + 2$ points. For example, 0 is the best constant approximation to $f(x) = \sin 4x$ on $[-\pi, \pi]$ and so $f - p_0^*$ has 16 different alternating sets consisting of 2 points. Indeed, the theorem tells us that $p_0^* = p_1^* = \cdots = p_5^* = p_6^* = 0$ are best approximations of indicated degree to $\sin 4x$ on $[-\pi, \pi]$ since $\max |\sin 4x| = 1$ and $\sin 4x$ "alternates" 8 times on $[-\pi, \pi]$, but $p_7 = 0$ is *not* a best approximation.

Theorem 1.7 also enables us to settle the uniqueness question for uniform approximation which was left open by Theorem I.3. We have

THEOREM 1.8. *If p_n^* is a best approximation to $f \in C[a, b]$ out of P_n, then it is unique; that is, if $p \in P_n$ and $p \neq p_n^*$,*

$$\|f - p\| > \|f - p_n^*\|.$$

Proof. Suppose $\|f - p\| = \|f - p_n^*\| = E_n(f)$ then, according to Theorem I.2,

$$q = \frac{p + p_n^*}{2}$$

is also a best approximation to f. Let $x_0, x_1, \ldots, x_{n+1}$ be an alternating set for $f - q$, so that for some integer l,

$$\frac{f(x_j) - p(x_j)}{2} + \frac{f(x_j) - p_n^*(x_j)}{2} = (-1)^{l+j} E_n(f), \qquad j = 0, \ldots, n+1.$$

Since

$$\frac{|f(x_j) - p(x_j)|}{2} \le \frac{E_n(f)}{2} \quad \text{and} \quad \frac{|f(x_j) - p_n^*(x_j)|}{2} \le \frac{E_n(f)}{2},$$

(1.2.10) can hold only if

$$f(x_j) - p(x_j) = f(x_j) - p_n^*(x_j) = (-1)^{l+j} E_n(f), \qquad j = 0, \ldots, n+1.$$

Thus, we obtain

$$p(x_j) = p_n^*(x_j), \qquad j = 0, \ldots, n+1,$$

which implies that $p = p_n^*$. The theorem is proved. ∎

Henceforth, we may refer to *the* best approximation. For example, the best approximating polynomial of degree 6 to $f(x) = \sin 4x$ on $[-\pi, \pi]$ is $p = 0$.

There are very few functions, $f(x)$, for which a best approximation, p_n^*, can be found explicitly. A most important instance in which this can be done is

$f(x) = x^{n+1}$. If p_n^* is the best approximation to x^{n+1} out of P_n and if $E_n(x^{n+1}; [-1, 1]) = \rho$, then

$$e(x) = x^{n+1} - p_n^*(x) \qquad (1.2.11)$$

satisfies

$$|e(x_j)| = \|e\| = \rho, \qquad j = 0, \ldots, n + 1, \qquad (1.2.12)$$

where $-1 \le x_0 < x_1 < \cdots < x_{n+1} \le 1$ in view of Theorem 1.7 and

$$e(x) \in P_{n+1}.$$

Now,

$$\rho^2 - e^2(x) \in P_{2n+2} \quad \text{and} \quad \rho^2 - e^2(x_j) = 0, \qquad j = 0, \ldots, n + 1.$$

Moreover, if $-1 < x_j < 1$, then since $e^2(x) \le \rho$ in $[-1, 1]$, $e^2(x)$ has a relative maximum at $x = x_j$ and so

$$\frac{d}{dx} e^2(x) = 0$$

for $x = x_j$. Thus at $x = x_j$

$$\frac{d}{dx} (\rho^2 - e^2(x)) = 0,$$

and each x_j in the interior of $[-1, 1]$ is a zero of multiplicity *two*, at least, of $\rho^2 - e^2(x)$. Since $\rho^2 - e^2(x)$ is not the zero polynomial, it has precisely $2n + 2$ zeros (counting multiple zeros as many times as their multiplicity); hence, *neither x_0 nor x_{n+1} are interior points of* $[-1, 1]$. That is, we have shown that $x_0 = -1$ and $x_{n+1} = 1$ are simple zeros while $x_1, \ldots, x_n$ are zeros of multiplicity two of $\rho^2 - e^2(x)$.

But the polynomial

$$(1 - x^2)[e'(x)]^2 \in P_{2n+2},$$

has simple zeros at ± 1 and has zeros of multiplicity two at $x_1, x_2, \ldots, x_n$. Thus $(1 - x^2)[e'(x)]^2$ and $\rho^2 - e^2(x)$ have identical sets of zeros and therefore, are constant multiples of one another. The constant is obtained by examining the leading coefficient of each, and we obtain

$$(1 - x^2)[e'(x)]^2 = (n + 1)^2(\rho^2 - e^2(x)). \qquad (1.2.13)$$

The differential equation (1.2.13) for the error function, $e(x)$, is easily solved. Since $e'(x) \in P_n$, and $e'(x_j) = 0$, $j = 1, \ldots, n$, we know that $e'(x)$ does not change sign in $[-1, x_1]$. We suppose $e'(x) \ge 0$ in $[-1, x_1]$, then (1.2.13) implies, for $x \in [-1, x_1]$,

$$\frac{e'(x)}{[\rho^2 - e^2(x)]^{1/2}} = \frac{(n + 1)}{(1 - x^2)^{1/2}}. \qquad (1.2.14)$$

Integrating both sides of (1.2.14) yields

$$\text{arc cos}\left(\frac{e(x)}{\rho}\right) = (n + 1)\theta + c,$$

where $x = \cos\theta$, $x \in [-1, x_1]$, $0 \le \theta_1 \le \theta \le \pi$. Thus we obtain

$$e(x) = \rho\cos[(n + 1)\theta + c].$$

Now $e(-1) = -\rho$ since we are assuming that $e'(-1) \ge 0$; thus, $\cos[(n + 1)\pi + c] = -1$ and $c = m\pi$, where $m + n + 1$ is odd. Hence,

$$e(x) = \pm\rho\cos(n + 1)\theta, \qquad (1.2.15)$$

$\cos(n + 1)\theta$ is a polynomial of degree $n + 1$ in $x = \cos\theta$; that is, $\cos(n + 1)\theta \in P_{n+1}$, and its leading coefficient is 2^n (cf. Exercise 1.6). (1.2.15) and (1.2.11) now imply that

$$e(x) = 2^{-n}\cos(n + 1)\theta. \qquad (1.2.16)$$

[It is clear that if $e'(x) \le 0$ in $[-1, x_1]$, we choose the negative square root on the right-hand side of (1.2.14), and the ensuing argument produces (1.2.16) again.]

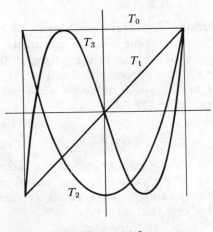

FIGURE 1.2

The polynomial $\cos k\theta$, where $x = \cos\theta$, $0 \le \theta \le \pi$, is called the *Chebyshev polynomial* of degree k and we write

$$T_k(x) = \cos k\theta, \qquad k = 0, 1, 2, \ldots,\dagger \qquad (1.2.17)$$

† The notation follows another transliteration from the Russian of "Chebyshev," one beginning with a T.

$T_k(x)$ is defined by (1.2.17) for $-1 \le x \le 1$ only. But because it is a polynomial, once we know its coefficients, we know it as a function for all x (complex numbers). A few of the Chebyshev polynomials are depicted in Figure 1.2. Thus, since $e(x) = 2^{-n}T_{n+1}(x)$ for $-1 \le x \le x_1$, the equality remains true for all x, in particular for $-1 \le x \le 1$. We collect this result as

THEOREM 1.9. *If p_n^* is a best approximation to x^{n+1} on $[-1, 1]$ out of P_n, then*

$$x^{n+1} - p_n^* = 2^{-n}T_{n+1}(x)$$

and so

$$E_n(x^{n+1}; [-1, 1]) = 2^{-n}.$$

The problem resolved by Theorem 1.9 is equivalent to finding that polynomial of degree k with leading coefficient 1 which deviates least from 0 in absolute value on $[-1, 1]$. Theorem 1.9 tells us that this polynomial is

$$\tilde{T}_k(x) = 2^{-k+1}T_k(x)$$

the *normalized* (that is, normalized so that its leading coefficient is 1) Chebyshev polynomial of degree k and, hence, that for each other polynomial of degree k with leading coefficient 1 there exists a point $x \in [-1, 1]$ such that

$$|p(x)| > \frac{1}{2^{k-1}}. \tag{1.2.18}$$

The Chebyshev polynomials have many interesting and useful properties, some of which are given in the exercises. We shall digress to present one of these. It may be interpreted as saying that since $T_n(x)$ uses up its oscillatory potentialities on $I: [-1, 1]$, it grows as rapidly as possible outside I. More precisely, we have

THEOREM 1.10. *If $p \in P_n$ and $|x_0| \ge 1$, then*

$$|p^{(k)}(x_0)| \le \|p\|\,|T_n^{(k)}(x_0)|, \qquad k = 0, 1, \ldots, n. \tag{1.2.19}$$

Proof. Suppose that the theorem is false, and there exists $y \ge 1$, say, such that

$$|p^{(k)}(y)| > \|p\|\,|T_n^{(k)}(y)| \tag{1.2.20}$$

for some $k = 0, 1, \ldots, n$. Put

$$q = \frac{p^{(k)}(y)}{|p^{(k)}(y)|} \cdot \frac{p}{\|p\|}, \tag{1.2.21}$$

so that $\|q\| = 1$. From (1.2.20), we see that $q \ne \pm T_n$. Let α be the leading coefficient of q; then $|\alpha| < 2^{n-1}$. For, if $\alpha = 0$, this is trivially true. If $\alpha \neq 0$, then q/α has leading coefficient 1. Hence, in view of (1.2.18),

$$\left\|\frac{q}{\alpha}\right\| = \frac{1}{|\alpha|} > \frac{1}{2^{n-1}}.$$

Let

$$\eta_j = \cos\frac{j\pi}{n}, \qquad j = 0, \ldots, n;$$

then

$$T_n(\eta_j) = (-1)^j, \qquad j = 0, \ldots, n.$$

For each $j = 0, \ldots, n - 1$, if $T_n(x) - q(x)$ does not have a zero at an end point of (η_{j+1}, η_j) (which is, necessarily, of multiplicity two or more for $j = 1, \ldots, n - 2$), then it has a zero in the interval (η_{j+1}, η_j). To see this, we observe that if $T_n(\eta_j) - q(\eta_j) \neq 0$ and $T_n(\eta_{j+1}) - q(\eta_{j+1}) \neq 0$, then $T_n(\eta_j) - q(\eta_j)$ and $T_n(\eta_{j+1}) - q(\eta_{j+1})$ have the same signs as $(-1)^j$ and $(-1)^{j+1}$, respectively. Thus $T_n(x) - q(x)$ changes sign in (η_{j+1}, η_j). If $T_n(\eta_j) - q(\eta_j) = 0$, $j = 1, \ldots, n - 1$, then since $\|q\| = 1$, $q'(\eta_j) = 0$ and, similarly, $T_n'(\eta_j) = 0$, so that η_j is a zero of multiplicity at least two. Thus we conclude that $T_n(x) - q(x)$ has all of its n zeros in $[-1, 1]$. Now, by Rolle's Theorem, $T_n^{(k)}(x) - q^{(k)}(x)$ has all $n - k$ of its zeros in $[-1, 1]$. Therefore, $T_n^{(k)}(x) - q^{(k)}(x)$ cannot change its sign for $x \geq 1$. But the coefficient of x^{n-k} in $T_n^{(k)}(x)$ is $n(n-1)\cdots(n-(k-1))2^{n-1}$ while the coefficient of x^{n-k} in $q^{(k)}(x)$ is $n(n-1)\cdots(n-(k-1))\alpha$. Thus,

$$T_n^{(k)}(x) - q^{(k)}(x) > 0$$

for x positive and sufficiently large and, in particular,

$$q^{(k)}(y) = \frac{|p^{(k)}(y)|}{\|p\|} \leq T_n^{(k)}(y) = |T_n^{(k)}(y)|,$$

contradicting (1.2.20) and thereby proving (1.2.19) for $x_0 \geq 1$. If $y \leq -1$ in (1.2.20), the argument is the same, word for word, if $n - k$ is even. If $n - k$ is odd, (1.2.21) must be replaced by

$$q = -\frac{p^{(k)}(y)}{|p^{(k)}(y)|} \cdot \frac{p}{\|p\|},$$

and then the same argument works. ∎

Remark. According to Theorem 1.10, $T_n(x)$ has the largest derivatives of all orders at $x = \pm 1$ among all $p \in P_n$ satisfying $\|p\| = 1$. As a matter of fact, even more is true. Namely, if $p \in P_n$ and $|p(\cos(j\pi)/n)| \leq 1, j = 0, \ldots, n$ (which is certainly the case if $\|p\| \leq 1$), then if $k \geq 1$,

$$\|p^{(k)}\| \leq T_n^{(k)}(1) = \frac{n^2(n^2-1)\cdots(n^2-(k-1)^2)}{1 \cdot 3 \cdot 5 \cdots (2k-1)}. \qquad (1.2.22)$$

This interesting result (in the case $\|p\| \leq 1$) is due to V. Markov [1],† but we cannot prove it here. The somewhat more general result stated here is due to Duffin and Schaeffer.

1.3   Approximation on a Finite Set of Points

Since it is usually not possible to find the polynomial of best uniform approximation to a given function, we are led to consider ways of approximating the best approximation. The procedure we shall consider is to replace the interval I by a finite set of points of I and solve the approximation problem on the finite point set. The success of this strategy depends on two factors: (1) the possibility of finding a best approximation to a function on a finite point set and (2) the best approximation on a finite point set approaching the best approximation on the interval as the number of points increases. We, therefore, turn next to a study of these two problems.

Let X_m be a set of m distinct points $x_1 < x_2 < \cdots < x_m$ of the interval $[a, b]$. If $m \leq n + 1$, there is a unique $p \in P_n$ satisfying

$$p(x_j) = f(x_j), \qquad j = 1, \ldots, m$$

for any f defined on X_m, and this p is the best approximation to f on X_m. Therefore, we suppose that $m \geq n + 2$. The best uniform approximation out of P_n to f on X_m is characterized by the exact analogue of Theorem 1.7. Since the proof of this fact is the same as that of Theorem 1.7 with X_m in place of $[a, b]$ everywhere, we omit it but state the result.

THEOREM 1.11. $p_n^*(X_m)$ *is a best uniform approximation on* X_m *to* f *out of* P_n *if and only if there exists an alternating set (of points of* X_m*) for* $f - p_n^*(X_m)$ *consisting of* $n + 2$ *points.*

The analogue of Theorem 1.8 now follows.

THEOREM 1.12.   *A best approximation on* X_m *is unique.*

We are now in a position to show that the approximation problem out of P_n on $[a, b]$ or on X_m can always be reduced to one on some X_{n+2}.

THEOREM 1.13.   *If* $p_n^* \in P_n$ *is the best approximation to* $f \in C[a, b]$*, there exists* $X_{n+2}^* \in [a, b]$ *such that*

$$E_n(f; [a, b]) = E_n(f; X_{n+2}^*) = \max_{x \in X_{n+2}^*} |f(x) - p_n^*(x)| < \max_{x \in X_{n+2}^*} |f(x) - p(x)|$$

† The case $k = 1$ is due to A. A. Markov.

for any $p \in P_n$, $p \neq p_n^*$. Moreover, for any $X_{n+2} \subset [a, b]$,

$$E_n(f; X_{n+2}) = \min_{p \in P_n} \max_{x \in X_{n+2}} |f(x) - p(x)|$$

$$= \max_{x \in X_{n+2}} |f(x) - p_n^*(X_{n+2}; x)| \leq E_n(f; [a, b]) = E_n(f; X_{n+2}^*)$$

$$(1.3.1)$$

with equality possible in (1.3.1) only if $p_n^*(X_{n+2}) = p_n^*$.

Proof. Let X_{n+2}^* be an alternating set for $f - p_n^*$, then, in view of Theorems 1.11 and 1.12 (with $X_m = X_{n+2}^*$), p_n^* is the best approximation to f on X_{n+2}^*, and the first part of the theorem is established.

Suppose $p_n^*(X_{n+2}) \neq p_n^*$; then, by uniqueness (Theorem 1.12),

$$E_n(f; X_{n+2}) < \max_{x \in X_{n+2}} |f(x) - p_n^*(x)| \leq E_n(f; [a, b]).$$

Equality can hold in (1.3.1) if $p_n^*(X_{n+2}) = p_n^*$, and X_{n+2} is an alternating set for $f - p_n^*$. ∎

We obtain, similarly, by replacing $[a, b]$ by X_m,

THEOREM 1.14.   *If $p_n^*(X_m) \in P_n$ is the best approximation to f on X_m, there exists $X_{n+2}^* \subseteq X_m$ such that*

$$E_n(f; X_m) = E_n(f; X_{n+2}^*) = \max_{x \in X_{n+2}^*} |f(x) - p_n^*(X_m; x)|$$

$$< \max_{x \in X_{n+2}^*} |f(x) - p(x)|$$

$$(1.3.2)$$

for any $p \in P_n$, $p \neq p_n^*(X_m)$. Moreover, for any $X_{n+2} \subseteq X_m$,

$$E_n(f; X_{n+2}) \leq E_n(f; X_{n+2}^*)$$

$$(1.3.3)$$

with equality possible only if $p_n^*(X_{n+2}) = p_n^*(X_m)$.

Theorems 1.13 and 1.14 enable us to reduce the search for a best approximation out of P_n to a set of $n + 2$ points. In the case of Theorem 1.13, this is not too helpful since there are infinitely many sets of $n + 2$ points in $[a, b]$. But in the case of Theorem 1.14 we see that, if we denote by $X_{m,i}$, $i = 1, \ldots, \binom{m}{n+2}$, the *finite* number of different subsets of $n + 2$ points of X_m, and find

$$\max_{i = 1, \ldots, \binom{m}{n+2}} E_n(f; X_{m,i}) = E_n(f; X_{m,i^*}),$$

then

$$p_n^*(X_m) = p_n^*(X_{m,i^*}).$$

This procedure presupposes our ability to find $E_n(f; X_{n+2})$ and $p_n^*(X_{n+2})$ for any set of $n + 2$ distinct points, X_{n+2}. Our next task is to see how this can be done.

Let us find the unique $p \in P_{n+1}$ which interpolates $f(x)$ at x_i, that is, such that

$$p(x_i) = f(x_i), \qquad i = 1, \ldots, n + 2. \tag{1.3.4}$$

This polynomial is given by the Lagrange interpolation formula as

$$p(x) = \sum_{i=1}^{n+2} \left\{ \prod_{j=1, j \neq i}^{n+2} \frac{(x - x_j)}{(x_i - x_j)} \right\} f(x_i), \tag{1.3.5}$$

as the reader may readily verify. If we put

$$\omega(x) = (x - x_1) \cdots (x - x_{n+2}),$$

(1.3.5) may be rewritten as

$$p(x) = \sum_{i=1}^{n+2} \frac{\omega(x)}{(x - x_i)} \cdot \frac{f(x_i)}{\omega'(x_i)}. \tag{1.3.6}$$

If

$$\sum_{i=1}^{n+2} \frac{f(x_i)}{\omega'(x_i)} = 0, \tag{1.3.7}$$

it follows that $p \in P_n$, since the left-hand side of (1.3.7) is the leading coefficient of p, and $p_n^*(X_{n+2}) = p$ while $E_n(f; X_{n+2}) = 0$, in view of (1.3.4). We suppose, therefore, that $p \notin P_n$, that is, the leading coefficient of p is not zero.

Let l_{n+1} be the unique polynomial in P_{n+1} which satisfies

$$l_{n+1}(x_i) = (-1)^i, \qquad i = 1, \ldots, n + 2.$$

That is,

$$l_{n+1}(x) = \sum_{i=1}^{n+2} \frac{\omega(x)}{(x - x_i)} \cdot \frac{(-1)^i}{\omega'(x_i)}. \tag{1.3.8}$$

We claim that $l_{n+1} \notin P_n$, that is, the leading coefficient of l_{n+1} is not zero. This leading coefficient is

$$\sum_{i=1}^{n+2} \frac{(-1)^i}{\omega'(x_i)}. \tag{1.3.9}$$

Now

$$\omega'(x_i) = \prod_{j=1, j \neq i}^{n+2} (x_i - x_j) = \left(\prod_{j=1}^{i-1} (x_i - x_j) \right) \left(\prod_{j=i+1}^{n+2} (x_j - x_i) \right) (-1)^{n+2-i},$$

so that

$$\operatorname{sgn} \omega'(x_i) = (-1)^{n-i} \dagger$$

---

† If x is any real (or complex) number, the signum of x, or sgn x, is defined by

$$\operatorname{sgn} x = \frac{x}{|x|}, \qquad x \neq 0, \qquad \operatorname{sgn} 0 = 0.$$

Thus for real x, sgn $x = 1$, $x > 0$; sgn $x = -1$, $x < 0$; sgn $0 = 0$.

and, hence,

$$\text{sgn} \frac{(-1)^i}{\omega'(x_i)} = (-1)^n.$$

All the terms (none of which is zero) in the sum (1.3.9) have the same sign. Thus the sum (1.3.9) cannot be zero, and our claim is established.

THEOREM 1.15

$$p_n^*(X_{n+2}) = p - \lambda l_{n+1},$$

where p is given by (1.3.6), l_{n+1} by (1.3.8), and

$$\lambda = \frac{\sum_{i=1}^{n+2} [f(x_i)/\omega'(x_i)]}{\sum_{i=1}^{n+2} [(-1)^i/\omega'(x_i)]}. \tag{1.3.10}$$

Moreover,

$$E_n(f; X_{n+2}) = |\lambda|. \tag{1.3.11}$$

Proof. We choose λ so that $p - \lambda l_{n+1} \in P_n$. (Note that l_{n+1} always has a nonzero leading coefficient while, if that of p is zero, then $\lambda = 0$.)
Now,

$$e(x_i) = f(x_i) - (p(x_i) - \lambda l_{n+1}(x_i)) = \lambda(-1)^i, \qquad i = 1, \ldots, n+2;$$

hence $x_1, \ldots, x_{n+2}$ form an alternating set for $f - (p - \lambda l_{n+1})$ on X_{n+2}, and by Theorem 1.11, $p_n^*(X_{n+2}) = p - \lambda l_{n+1}$. Hence, (1.3.11) holds, as well. ∎

Theorem 1.15 provides a complete theoretical solution to the problem of uniform approximation out of P_n on $n + 2$ points. If we are given a finite point set X_m, then we use (1.3.10) to compute $|\lambda|$ for each subset of X_m consisting of $n + 2$ points, pick out a largest $|\lambda|$ from the finite sequence so computed, and the p_n^* of the subset associated with the largest $|\lambda|$ is the best approximation out of P_n on X_m. We illustrate with the example $X_m = \{-1, -\frac{1}{2}, 0, \frac{1}{2}, 1\}$, so that $m = 5$, and find the best approximation of degree at most 2 to $f(x) = |x|$. The reader should verify that if

$$X_{5,1} = \{-1, -\tfrac{1}{2}, 0, \tfrac{1}{2}\}; \qquad X_{5,2} = \{-1, -\tfrac{1}{2}, 0, 1\};$$

$$X_{5,3} = \{-1, -\tfrac{1}{2}, \tfrac{1}{2}, 1\}; \qquad X_{5,4} = \{-1, 0, \tfrac{1}{2}, 1\};$$

$$X_{5,5} = \{-\tfrac{1}{2}, 0, \tfrac{1}{2}, 1\},$$

in obvious notation $|\lambda^{(1)}| = \frac{1}{8}$, $|\lambda^{(2)}| = \frac{1}{9}$, $|\lambda^{(3)}| = 0$, $|\lambda^{(4)}| = \frac{1}{9}$, $|\lambda^{(5)}| = \frac{1}{8}$, and $p_2^*(X_m) = x^2 + \frac{1}{8}$. (As a matter of fact, it is easy to see that $p_2^* = x^2 + \frac{1}{8}$ and $X_{5,1}$ and $X_{5,5}$ are both alternating sets for $|x| - p_2^*$. The fact that $|x|$ and p_2^* are both even functions jibes with Exercise 1.1.)

While this technique is entirely satisfactory from a theoretical point of view, in practice it may be forbiddingly lengthy when m is substantially larger than n. What is needed is a strategy for choosing the subsets $X_{n+2,i}$ of X_m instead of proceeding blindly through all of them. Before we examine a computational strategy, we shall examine the relationship between best approximation on an interval and best approximation on a mesh of points of the interval. It is vital to have this information in order to determine whether the intractable interval-approximation problem can be replaced by a much easier finite-point-set approximation problem, and if so, to determine how many points are needed to obtain a prescribed accuracy.

We normalize the interval to $I : [-1, 1]$ and suppose $X_m \subset I$ is a set of m distinct points, including the points $x = -1, 1$. Let $\delta_m > 0$ be a number such that for each $x \in I$ there exists $x_i \in X_m$ and

$$|x - x_i| \leq \delta_m;$$

that is, δ_m is the greatest distance from a point of I to X_m. If, for example, X_m consists of equally spaced points in I, that is, the points

$$-1 + \frac{2j}{m-1}, \quad j = 0, \ldots, m-1, \tag{1.3.12}$$

then

$$\delta_m = \frac{1}{m-1}. \tag{1.3.13}$$

For this choice of points δ_m, clearly, has its smallest value.

We need some preliminary results. The first is due to Ehlich and Zeller [1].

LEMMA 1.8. *If $p \in P_n$ satisfies*

$$|p(x)| \leq K, \quad x \in X_m \tag{1.3.14}$$

and

$$\delta_m < \frac{\sqrt{6}}{n\sqrt{n^2 - 1}}, \tag{1.3.15}$$

then

$$|p(x)| \leq \frac{K}{1 - \tau}, \quad x \in I, \tag{1.3.16}$$

where

$$\tau = \frac{n^2(n^2 - 1)}{6} \delta_m^2. \tag{1.3.17}$$

Proof. (1.3.15) implies that $\tau < 1$. Suppose that $|p(\xi)| = \|p\|$. If $\xi = \pm 1$, then $\xi \in X_m$ and (1.3.16) follows trivially from (1.3.14). Therefore, we suppose that $-1 < \xi < 1$ and let x_i be the point of X_m nearest to ξ. Then

$$p(x_i) = p(\xi) + (x_i - \xi)p'(\xi) + \frac{(x_i - \xi)^2}{2!} p''(\eta)$$

for some $\eta \in I$. But $p'(\xi) = 0$ since $p(x)$ has a relative extremum at ξ; hence

$$\|p\| = |p(\xi)| \leq K + \frac{\delta_m^2}{2}|p''(\eta)|. \tag{1.3.18}$$

We need an upper bound for $|p''(\eta)|$. This is provided by the result of V. Markov (1.2.22) with $k = 2$:

$$|p''(\eta)| \leq \frac{n^2(n^2 - 1)}{3}\|p\|. \tag{1.3.19}$$

Substituting (1.3.19) into (1.3.18) yields

$$\|p\| \leq K + \frac{\delta_m^2}{6}n^2(n^2 - 1)\|p\| = K + \tau\|p\|,$$

from which (1.3.16) follows. ∎

LEMMA 1.9. *For $p \in P_n$*

$$\omega(p; I; \delta) \leq \delta n^2 \|p\|. \tag{1.3.20}$$

Proof. If $x', x'' \in I$, then, by the mean-value theorem,

$$p(x') - p(x'') = (x' - x'')p'(\eta).$$

But, relying on Markov's result (1.2.22),

$$|p'(\eta)| \leq n^2\|p\|$$

and, hence,

$$|p(x') - p(x'')| \leq \delta n^2 \|p\|,$$

which implies (1.3.20). ∎

THEOREM 1.16. *If $f(x)$ is continuous on I, then*

$$E_n(f; I) - \omega(f; I; \delta_m) - \delta_m n^2 \left[\frac{\|f\| + E_n(f; I)}{1 - \tau}\right] \leq E_n(f; X_m) \leq E_n(f; I), \tag{1.3.21}$$

provided that δ_m satisfies (1.3.15).

Proof. Let p_n^* be the best approximation to f on I and q_n^* the best approximation to f on X_m. Then

$$E_n(f; X_m) \leq \max_{x \in X_m}|f(x) - p_n^*(x)| \leq \max_{x \in I}|f(x) - p_n^*(x)| = E_n(f; I), \tag{1.3.22}$$

and the right-hand inequality in (1.3.21) is proved.

Given $x \in I$, let $x_i \in X_m$ be the point of X_m nearest to x; then

$$f(x) - q_n^*(x) = [f(x) - f(x_i)] + [f(x_i) - q_n^*(x_i)] + [q_n^*(x_i) - q_n^*(x)],$$

and hence

$$|f(x) - q_n^*(x)| \leq \omega(f; I; \delta_m) + E_n(f; X_m) + \omega(q_n^*; I; \delta_m). \quad (1.3.23)$$

Since $E_n(f; I) \leq \|f - q_n^*\|$, we obtain from (1.3.23)

$$E_n(f; X_m) \geq E_n(f; I) - \omega(f; I; \delta_m) - \omega(q_n^*; I; \delta_m). \quad (1.3.24)$$

The theorem will be proved by next finding an upper bound for $\omega(q_n^*; I; \delta_m)$. By combining the results of Lemmas 1.9 and 1.8, we obtain

$$\omega(q_n^*; I; \delta_m) \leq \delta_m n^2 \|q_n^*\| \leq \delta_m n^2 (1 - \tau)^{-1} \max_{x \in X_m} |q_n^*(x)|. \quad (1.3.25)$$

But since $|f(x) - q_n^*(x)| \leq E_n(f; X_m)$ for $x \in X_m$,

$$\max_{x \in X_m} |q_n^*(x)| \leq E_n(f; X_m) + \|f\| \leq E_n(f; I) + \|f\|. \quad (1.3.26)$$

(1.3.24), (1.3.25), and (1.3.26) combine to prove the theorem. ∎

COROLLARY 1.16.1.   *If $\delta_m \to 0$ as $m \to \infty$, then $E_n(f; X_m) \to E_n(f; I)$ as $m \to \infty$.*

Proof.   As $m \to \infty$, $\omega(f; I; \delta_m) \to 0$ since f is continuous on I (Lemma 1.2). Since $\tau \to 0$, (1.3.21) now implies the corollary. ∎

Suppose that X_m consists of the points (1.3.12). We are now in a position to answer the question, how large does m have to be in order that $E_n(f; X_m)$ be within ε of $E_n(f; I)$? The answer is implicit in (1.3.21), and depends on how smooth f is. A typical result is

COROLLARY 1.16.2.   *Given $f \in \text{lip}_M 1$ on I and $\varepsilon > 0$, then if*

$$m > \max\left(1 - n^2, 1 + \frac{M + \frac{6}{5}n^2(\|f\| + 6Mn^{-1})}{\varepsilon}\right), \quad (1.3.27)$$

we have, in the case of equally spaced points,

$$E_n(f; I) - \varepsilon < E_n(f; X_m) \leq E_n(f; I). \quad (1.3.28)$$

Proof.   In view of (1.3.21) and (1.1.28), it is enough to show that for some λ satisfying $0 < \lambda < 1$

$$\frac{M}{(m - 1)} < \lambda\varepsilon \quad (1.3.29)$$

and

$$\frac{n^2}{m - 1}\left[\frac{\|f\| + 6Mn^{-1}}{1 - (n^4/6(m - 1)^2)}\right] < (1 - \lambda)\varepsilon. \quad (1.3.30)$$

Since $m > n^2 + 1$, $1 - n^4/(6(m - 1)^2) > \frac{5}{6}$, and (1.3.30) will follow from

$$\frac{1}{m - 1}\frac{6}{5}n^2[\|f\| + 6Mn^{-1}] < (1 - \lambda)\varepsilon. \tag{1.3.31}$$

(1.3.29) and (1.3.31) are equivalent to

$$m > 1 + \frac{M}{\lambda\varepsilon} \tag{1.3.32}$$

and

$$m > 1 + \frac{\frac{6}{5}n^2[\|f\| + 6Mn^{-1}]}{(1 - \lambda)\varepsilon}, \tag{1.3.33}$$

respectively. We now choose λ so that the lower bounds on m in (1.3.32) and (1.3.33) are equal:

$$\lambda = \frac{M}{M + \frac{6}{5}n^2[\|f\| + 6Mn^{-1}]}. \tag{1.3.34}$$

The proof is concluded by substituting (1.3.34) in (1.3.32). ∎

The reader may, if he wishes, easily find other variations on the theme of this corollary. A full discussion will be found in Rivlin and Cheney [1].

As an example of Corollary 1.16.2 suppose that $f(x) = |x|$. Since $||x'| - |x''|| \le |x' - x''|$, we see that $f \in \text{lip}_1 1$. Hence (1.3.27) becomes

$$m > \max\left[1 + n^2, 1 + \frac{1 + \frac{6}{5}n^2(1 + (6/n))}{\varepsilon}\right] = 1 + \frac{1 + \frac{6}{5}n^2(1 + (6/n))}{\varepsilon}$$

for ε sufficiently small. When $n < 12$, a better bound for $E_n(f; I)$ than $6/n$ is $\frac{1}{2}$ (that this *is* a bound is clear from a diagram). In particular, if (as in the example worked out on p. 36) $n = 2$, we obtain

$$m > 1 + \frac{41}{5\varepsilon},$$

a safe but overly pessimistic bound.

1.4 Computational Methods

1.4.1 The Exchange Method

We turn next to a strategy for choosing subsets of $n + 2$ points out of X_m. We call a subset of $n + 2$ points of X_m (arranged on the line from left to right) a *reference* and denote the points of a reference by x_σ (Greek subscripts will always run over $n + 2$ indices among $1, \ldots, m$). We write $\{x_\sigma\}$ for the reference itself.

Suppose that $\{x_\sigma\}$ is a reference. Let $p_\sigma = p_n^*(\{x_\sigma\})$ be the polynomial of degree n of best approximation to f on $\{x_\sigma\}$ that is determined by Theorem 1.15. Then

$$|f(x_\sigma) - p_\sigma(x_\sigma)| = E_n(f; \{x_\sigma\}),$$

where $E_n(f; \{x_\sigma\})$ is determined by (1.3.10) and (1.3.11). Let us write $\rho_\sigma = E_n(f; \{x_\sigma\})$ and put

$$M_\sigma = \max_{x \in X_m} |f(x) - p_\sigma(x)| = |f(x_j) - p_\sigma(x_j)|.$$

There are two possibilities: (i) $M_\sigma = \rho_\sigma$ or (ii) $M_\sigma > \rho_\sigma$; hence, $x_j \notin \{x_\sigma\}$. In the former case, for any $p \in P_n$, $p \neq p_\sigma$,

$$\max_{x \in X_m} |f(x) - p_\sigma(x)| = \rho_\sigma < \max |f(x_\sigma) - p(x_\sigma)| \leq \max_{x \in X_m} |f(x) - p(x)|$$

and $p_\sigma = p_n^*(X_m)$. Thus, in this case our quest is ended, and $\{x_\sigma\}$ is an alternating set for $f - p_n^*(X_m)$. Let us suppose then that $M_\sigma > \rho_\sigma$. We now *exchange* x_j for one of the points of the reference $\{x_\sigma\}$ in order to form a new reference $\{x_\mu\}$. The point of $\{x_\sigma\}$ which x_j replaces is chosen so that $f(x) - p_\sigma(x)$ alternates in sign as x traverses $\{x_\mu\}$. If x_j lies between two consecutive points of $\{x_\sigma\}$, it replaces the one of its two neighbors, x_σ, for which $f(x_\sigma) - p_\sigma(x_\sigma)$ has the same sign as $f(x_j) - p_\sigma(x_j)$ (remember that $f - p_\sigma$ alternates in sign on $\{x_\sigma\}$). However, if x_j lies to the right of all points of $\{x_\sigma\}$, it replaces the rightmost point of $\{x_\sigma\}$ when $f - p_\sigma$ has the same sign at this point as at x_j; the left-most point is deleted from $\{x_\sigma\}$ when the signs disagree. A similar argument can be given when x_j lies to the left of $\{x_\sigma\}$. Using Theorem 1.15, we now determine p_μ and ρ_μ. The point of the method is contained in the inequality

$$\rho_\sigma < \rho_\mu.$$

(To see the importance of this inequality, recall that we are seeking the subset $\{x_\nu\}$ for which the corresponding ρ_ν is *greatest*.) To prove the inequality, we note that taking $X_{n+2} = \{x_\mu\}$ in Exercises 1.23 and 1.24 results in

$$\pm \rho_\mu = \sum (-1)^\mu \lambda_\mu (f(x_\mu) - p_\sigma(x_\mu)), \qquad (1.4.1)$$

where

$$\lambda_\mu > 0, \qquad \sum \lambda_\mu = 1. \qquad (1.4.2)$$

But since $f(x_\mu) - p_\sigma(x_\mu)$ alternates in sign on $\{x_\mu\}$, (1.4.1) implies that

$$\rho_\mu = \sum_{\mu \neq j} \lambda_\mu \rho_\sigma + \lambda_j M_\sigma = \sum_\mu \lambda_\mu \rho_\sigma + \lambda_j (M_\sigma - \rho_\sigma) = \rho_\sigma + \lambda_j (M_\sigma - \rho_\sigma) > \rho_\sigma$$

in view of (1.4.2) and the positivity of $M_\sigma - \rho_\sigma$.

Next we calculate M_μ. If $M_\mu = \rho_\mu$, the process terminates and $p_\mu = p_n^*(X_m)$.

If $M_\mu > \rho_\mu$, we repeat the exchange procedure to obtain a new reference $\{x_\alpha\}$ such that $\rho_\alpha > \rho_\mu$. At each step we get a new reference, and since X_m contains only finitely many references, the process terminates in a finite number of steps. Thus the exchange method is, in general, a more sensible strategy than proceeding through all the references $X_{n+2,i}$ of X_m in a random fashion since the reference deviations increase monotonically.

The question of which reference to choose at the start arises. A practical answer can be given by jumping ahead to our chapter on least-squares approximation (Chapter 2). If q_n is the best least-squares approximation of degree at most n to f on X_m, that is,

$$\sum_{x \in X_m} [f(x) - q_n(x)]^2 < \sum_{x \in X_m} [f(x) - p(x)]^2$$

for any $p \in P_n$, $p \neq q_n$, then as we shall see, there exist $n + 2$ points of X_m on which $f - q_n$ alternates in sign. These points are a good choice for a starting reference. (It will also turn out that q_n can always be found quite easily, so we have not just exchanged one difficulty for another of equal magnitude.) Let us examine the example we worked out on p. 36, that is, approximation by quadratics to $|x|$ on X_5. The least-squares approximation on X_5 turns out to be $q_2 = \frac{6}{7}x^2 + \frac{6}{35}$, and $|x| - q_2$ alternates in sign on $X_{5,1}$ and $X_{5,5}$, so that we are led at once to the optimal references. Suppose, however, that we start with $X_{5,2} = \{-1, -\frac{1}{2}, 0, 1\}$ as a reference, σ. Then $p_\sigma = \frac{8}{9}x^2 - \frac{1}{9}x + \frac{1}{9}$, $M_\sigma = \frac{2}{9}$, and $x_j = \frac{1}{2}$. $|x| - p_\sigma = -\frac{1}{9}$ at $x = 0$, $\frac{2}{9}$ at $x = \frac{1}{2}$, and $\frac{1}{9}$ at $x = 1$. We therefore exchange the point 1 of the reference for the point $\frac{1}{2}$, obtaining a new reference, namely $X_{5,1}$, which we know to be optimal. The point is that we were not led to $X_{5,3}$. For this and other variants of the exchange method, we refer the reader to the following literature: Rice [1], Meinardus [1], Remez [1], Stiefel [1].

1.4.2 Linear Programming

Another approach to the problem of finding best approximations on X_m is to express the problem as a linear programming problem. In addition to the $n + 1$ unknown coefficients of $p_n^*(X_m)$, we introduce a new variable, ρ, as follows. The condition

$$\max_{x \in X_m} |f(x) - p(x)| = \rho$$

can be written

$$-\rho \leq f(x_j) - \sum_{i=0}^{n} a_i x_j^i \leq \rho, \qquad j = 1, \ldots, m.$$

Our problem then is to *minimize* the linear function of $a_0, a_1, \ldots, a_n, \rho$,

$$\rho,$$

subject to the $2m$ linear constraints

$$\rho + \sum_{i=0}^{n} a_i x_j^i \geq f(x_j), \qquad j = 1, \ldots, m,$$

$$\rho - \sum_{i=0}^{n} a_i x_j^i \geq -f(x_j), \qquad j = 1, \ldots, m.$$

Either the problem can be easily "adjusted" to become a primal problem, and then the simplex method applied, or the dual-simplex method can be used directly. We shall not enter into these techniques here but refer the reader to the general literature (cf, Dantzig [1]), or better still to a detailed discussion including the relationship between the simplex method and the exchange and other methods in Stiefel [1], and the improvements of Barrodale and Young [1]. See also Rabinowitz [1].

Exercises

1.1 If $f(x)$ is an even (odd) function on $[-a, a]$ that has a best approximation, show that it has a best approximation that is also even (odd).

1.2 If $[c, d]$ is a subinterval of $[a, b]$, show that

$$\omega(f; [c, d]; \delta) \leq \omega(f; [a, b]; \delta).$$

1.3 If f is an even function on $[-a, a]$, show that

$$\omega(f; [-a, a]; \delta) = \omega(f; [0, a]; \delta).$$

[*Hint:* Suppose that $x_1, x_2 \in [-a, a]$, $x_1 \geq 0$, $x_2 < 0$, $|x_1 - x_2| \leq \delta$, then $|f(x_1) - f(x_2)| = |f(x_1) - f(-x_2)|$ and $|x_1 + x_2| \leq \delta$.]

1.4 If $x = \cos \theta$, $-1 \leq x \leq 1$, and $g(\theta) = f(\cos \theta)$, show that

$$\omega(g; [-\pi, \pi]; \delta) = \omega(g; [0, \pi]; \delta) \leq \omega(f; [-1, 1]; \delta).$$

1.5 If $f(x)$ has period T, show that

$$\omega(f; [a, a + T]; \delta)$$

is independent of a.

1.6 Show that $\cos k\theta$ is a polynomial of degree k in $\cos \theta$. What is the coefficient of $(\cos \theta)^k$?

[*Hint:* $\cos k\theta = \text{Re} (\cos \theta + i \sin \theta)^k$.]

1.7 Suppose that $g(x) = f(Ax + B)$ for $c \leq x \leq d$; show that

$$\omega(g; [c, d]; \delta) = \omega(f; [Ac + B, Ad + B]; A\delta).$$

1.8   If $|f'(x)| \le M$ on $c \le x \le d$, show that

$$\omega(f; [c, d]; \delta) \le M\delta.$$

1.9   Show that $E_n(f; [a, b]) = E_n(f - p; [a, b])$ for all $p \in P_n$. If p_n^* is a best approximation to f out of P_n, then $p_n^* - p$ is a best approximation to $f - p$ out of P_n, for $p \in P_n$.

1.10   If $f''(x) > 0$ for all $x \in [a, b]$ (so that f is convex), show that a best approximation of degree at most *one* to $f(x)$ is

$$p_1^*(x) = \frac{f(a)(b - c) + f(b)(c - a) + f(c)(b - a)}{2(b - a)} + \frac{f(b) - f(a)}{b - a}(x - c)$$

and

$$E_1(f; [a, b]) = \frac{f(a)(b - c) + f(b)(c - a) - f(c)(b - a)}{2(b - a)},$$

where c is the unique solution (in $[a, b]$) of $f'(c) = (f(b) - f(a))/(b - a)$.

1.11   Show that the best linear approximation to $f(x) = (1 + x^2)^{1/2}$ on $[0, 1]$ is $(2c + 1)/2 + (\sqrt{2} - 1)x$, where $c = ((\sqrt{2} - 1)/2)^{1/2}$. Thus, we obtain the Pythagorean approximation: $\sqrt{u^2 + v^2}$ is approximated by $[(2c + 1)/2]u + (\sqrt{2} - 1)v$ if $u \ge v$. What is the error in this approximation?

1.12   If $f^{(n+1)}(x)$ exists and is never zero in $[a, b]$ and p_n^* is a best approximation to f, show that $f(x) - p_n^*(x)$ has a unique alternating set $x_0, \ldots, x_{n+1}$ and $x_0 = -1$, $x_{n+1} = 1$.

1.13   Suppose that $q \in P_n$, $f \in C[a, b]$, and

$$f(x_j) - q(x_j) = (-1)^j A_j$$

with $A_j > 0$, $j = 0, \ldots, n + 1$. Prove that

$$E_n(f) \ge \min (A_0, \ldots, A_{n+1}).$$

1.14   Show that $T_k(x)$ is even for even k and odd for odd k.

1.15   Find the best approximation to x^{n+2} out of P_n.

1.16   Verify that the Chebyshev polynomials have the following properties:

(a) $$(1 - x^2)T_n''(x) - xT_n'(x) + n^2 T_n(x) = 0.$$

[Use this result to verify the expression for $T_n^{(k)}(1)$ given in (1.2.22).]

(b) Three-term recurrence

$$T_n(x) = 2xT_{n-1}(x) - T_{n-2}(x), \qquad n = 2, 3, \ldots.$$

(c) Orthogonality

$$\int_{-1}^{1} T_k(x)T_m(x) \frac{dx}{\sqrt{1 - x^2}} = 0, \qquad m \ne k,$$

$$\int_{-1}^{1} T_k^2(x) \frac{dx}{\sqrt{1 - x^2}} = \begin{cases} \pi/2, & k > 0, \\ \pi, & k = 0. \end{cases}$$

(d) $\frac{1}{2}T_k(1)T_m(1) + \displaystyle\sum_{j=1}^{n-1} T_k(\eta_j)T_m(\eta_j) + \frac{1}{2}T_k(-1)T_m(-1)$

$$= \begin{cases} 0, & k \neq m, \\ n/2, & k = m \neq 0, n, \\ n, & k = m = 0, n, \end{cases}$$

when $m, k \leq n$, and the η_j are the extrema of T_n as defined on p. 32.

(e) Semi-group property

$$T_m(T_n(x)) = T_{mn}(x); \qquad m, n > 0.$$

(f) $\qquad T_n(x) = \dfrac{(x + \sqrt{x^2 - 1})^n + (x - \sqrt{x^2 - 1})^n}{2}.$

(g) If we write

$$T_n(x) = \sum_{j=0}^{n} t_j^{(n)} x^j,$$

then

$$t_{n-2m}^{(n)} = (-1)^m \frac{n}{n-m} \binom{n-m}{m} 2^{n-2m-1}, \qquad m = 0, 1, 2, \ldots, \left[\frac{n}{2}\right],$$

and all other coefficients are zero.

1.17 Show that on $[a, b]$ the polynomial of degree k with leading coefficient 1 that has the least absolute deviation from 0 is

$$\tilde{T}_n\left(\frac{2}{b-a}x - \frac{b+a}{b-a}\right).$$

1.18 Prove that:

(i) If $p \in P_n$, $\max_{x \in I} |p(x)| = 1$, and there exist $(n+1)$ distinct points $x_i \in I$ such that $|p(x_i)| = 1$, $i = 0, \ldots, n$, then $p(x) = \pm T_n(x)$, or $p = \pm 1$.

(ii) If $t \in \mathcal{T}_n$, $\max_{0 \leq \theta \leq 2\pi} |t(\theta)| = 1$, and there exist $2n$ distinct points θ_i satisfying $0 \leq \theta_i < 2\pi$ and $|t(\theta_i)| = 1$, $i = 1, \ldots, 2n$, then

$$t(\theta) = \cos(n\theta + \theta_0).$$

1.19 Show that every $p \in P_n$ has a unique representation in the form

$$p(x) = A_0 + A_1 T_1(x) + \cdots + A_n T_n(x).$$

1.20 Show that the best approximation out of P_n to $1/(x-a)$, $a > 1$, on $-1 \leq x \leq 1$ is

$$p_n^*(x) = \frac{-2t}{t^2 - 1} + \frac{4t}{t^2 - 1} \sum_{j=0}^{n-1} t^j T_j(x) - \frac{4t^{n+1}}{(1-t^2)^2} T_n(x),$$

where $t = a - (a^2 - 1)^{1/2}$. Use this result together with Exercise 1.9 to find the best approximation out of P_n on $[-1, 1]$ to $q(x)/(x - a)$ for any $q \in P_{n+1}$.

[*Hint:* (Rivlin [1]) With $x = \cos \theta$, it is possible to sum the series

$$\sum_{j=0}^{\infty} t^j T_j(x) = \sum_{j=0}^{\infty} t^j \cos j\theta = \text{Re} \sum_{j=0}^{\infty} (t\, e^{i\theta})^j.$$

For the second part, find $p \in P_n$ such that

$$\frac{q(x)}{x - a} = \frac{c}{x - a} - p.]$$

1.21   Show that

$$T_n^{(k)}(x) > 0 \qquad \text{for} \qquad x \geq 1, \qquad k = 0, \ldots, n$$

and

$$\text{sgn}\, T_n^{(k)}(x) = (-1)^n \qquad \text{for} \qquad x \leq -1, \qquad k = 0, \ldots, n.$$

[*Hint:* Use Rolle's Theorem.]

1.22   Determine the conditions under which there can be equality in (1.2.19), (Theorem 1.10).

1.23   Verify that (1.3.10) is equivalent to

$$\lambda = \sum_{i=1}^{n+2} (-1)^i \lambda_i f(x_i),$$

where

$$\lambda_i = \frac{1/|\omega'(x_i)|}{\sum_{i=1}^{n+2} (1/|\omega'(x_i)|)}, \qquad i = 1, \ldots, n + 2,$$

and hence

$$\lambda_i > 0 \qquad \text{and} \qquad \sum_{i=1}^{n+2} \lambda_i = 1.$$

1.24   Show that, in the notation of Exercise 1.23,

$$\sum_{i=1}^{n+2} (-1)^i \lambda_i q(x_i) = 0$$

for any $q \in P_n$.

1.25   (L. Smith) Given $f \in C[-1, 1]$, consider the problem of finding $\bar{p} \in P_n$ with the property that

$$f(x) \geq \bar{p}(x), \qquad -1 \leq x \leq 1,$$

and

$$\max_{-1 \leq x \leq 1} [f(x) - \bar{p}(x)] \leq \max_{-1 \leq x \leq 1} [f(x) - p(x)]$$

for all $p \in P_n$ satisfying $f(x) \geq p(x)$, $-1 \leq x \leq 1$. (Problem of *one-sided* approximation from below.)

Show that, if p^* is the best uniform approximation to f on $[-1, 1]$ out of P_n, then $\bar{p} = p^* - E_n(f; [-1, 1])$. Consider, similarly, the problem of one-sided approximation from above. Study the uniqueness and characterization problems for one-sided approximation.

Notes, Chapter 1

(1) The proof presented here is taken from Korovkin [1]. In his book Korovkin presents a generalization of the method used here with many examples.

(2) The proof of Jackson's Theorem which we have given, follows Korovkin [1]. (See also Meinardus [1].) Jackson's original proof is somewhat different. See, for example, Jackson [1].

(3) This theorem and other results in this and other chapters continue to hold when the approximating family is spanned by a Chebyshev system, a generalization of the monomials $1, x, \ldots, x^n$ which span P_n. A Chebyshev system is a set of continuous functions $u_0(x), \ldots, u_n(x)$ on $[a, b]$ with the property that every linear combination ("polynomial") $a_0 u_0(x) + \cdots + a_n u_n(x)$ has at most n distinct zeros in $[a, b]$ unless $a_0 = \ldots = a_n = 0$. The properties of such systems, and interesting examples of them, are to be found in Karlin and Studden [1]. As an instructive project the reader should try to determine which results in this chapter, and indeed in subsequent chapters, extend to Chebyshev systems. Karlin and Studden [1] is a valuable guide in such an undertaking.

CHAPTER 2

LEAST-SQUARES APPROXIMATION

We begin by studying least-squares polynomial approximation on an interval. First we characterize the least-squares approximation and then demonstrate the efficacy of orthogonal polynomials in constructing it. There follows a more detailed discussion of perhaps the most useful sets of orthogonal polynomials, the Legendre polynomials, and our old friends, the Chebyshev polynomials. The next topic is least-squares approximation on a finite point set, and the final section considers the utility of least-squares approximations as uniform approximations.

2.1 Approximation on an Interval

It is a consequence of Theorems I.1 and I.3 that, given a continuous function $f(x)$ on I: $[-1, 1]$ and a (Riemann) integrable *weight function* $w(x)$, which is positive on I (except, possibly, at a finite number of points of I at which $w(x) = 0$), there exists a unique $q_n^* \in P_n$ such that

$$\|f - q_n^*\|_2 = \left[\int_{-1}^{1} [f(x) - q_n^*(x)]^2 w(x) \, dx \right]^{1/2} < \|f - p\|_2 \quad (2.1.1)$$

for any $p \in P_n$, $p \neq q_n^*$. This q_n^* is called the least-squares approximation to f out of P_n (with respect to the weight function $w(x)$).

In contrast to the best uniform approximation, we shall see shortly that q_n^* is rather easily obtainable. We begin by characterizing the least-squares approximation.

THEOREM 2.1.   *Given $f(x) \in C(I)$ and a weight function on I, $w(x)$, then q_n^* is the least-squares approximation to f out of P_n if and only if*

$$\int_{-1}^{1} (f(x) - q_n^*(x)) p(x) w(x) \, dx = 0 \quad (2.1.2)$$

for every $p \in P_n$.

Proof. (i) Suppose (2.1.2) holds. Then, if $p \in P_n$,

$$\int_{-1}^{1} (f(x) - p(x))^2 w(x)\, dx = \int_{-1}^{1} [(f(x) - q_n^*(x)) + (q_n^*(x) - p(x))]^2 w(x)\, dx$$

$$= \int_{-1}^{1} (f(x) - q_n^*(x))^2 w(x)\, dx$$

$$+ 2 \int_{-1}^{1} (f(x) - q_n^*(x))(q_n^*(x) - p(x)) w(x)\, dx$$

$$+ \int_{-1}^{1} (q_n^*(x) - p(x))^2 w(x)\, dx. \tag{2.1.3}$$

Since $q_n^* - p \in P_n$, (2.1.2) implies that the second term on the right in (2.1.3) is zero. Thus

$$\int_{-1}^{1} (f(x) - q_n^*(x))^2 w(x)\, dx \le \int_{-1}^{1} (f(x) - p(x))^2 w(x)\, dx$$

and hence

$$\|f - q_n^*\|_2 \le \|f - p\|_2;$$

that is, q_n^* is the least-squares approximation.

(ii) Suppose there exists $\bar{p} \in P_n$ such that

$$\int_{-1}^{1} (f(x) - q_n^*(x)) \bar{p}(x) w(x)\, dx = a \ne 0.$$

Then

$$b = \int_{-1}^{1} (\bar{p}(x))^2 w(x)\, dx > 0,$$

and if we put

$$\lambda = \frac{a}{b} \ne 0,$$

$$\int_{-1}^{1} (f(x) - q_n^*(x) - \lambda \bar{p}(x))^2\, w(x)\, dx = \int_{-1}^{1} (f(x) - q_n^*(x))^2 w(x)\, dx - 2\lambda a + \lambda^2 b$$

$$= \int_{-1}^{1} (f(x) - q_n^*(x))^2 w(x)\, dx - \lambda^2 b.$$

But since $\lambda^2 b > 0$, we conclude that

$$\|f - (q_n^* + \lambda \bar{p})\|_2 < \|f - q_n^*\|_2,$$

and q_n^* is not the least-squares approximation to f. ∎

If we put $p = x^i$, $i = 0, 1, \ldots, n$ successively in (2.1.2), we obtain

$$\int_{-1}^{1} x^i q_n^*(x) w(x)\, dx = \int_{-1}^{1} x^i f(x) w(x)\, dx, \qquad i = 0, \ldots, n, \tag{2.1.4}$$

a system of $(n + 1)$ linear equations for the $(n + 1)$ unknown coefficients of $q_n^*(x)$. If we write

$$q_n^*(x) = \xi_0 + \xi_1 x + \cdots + \xi_n x^n,$$

the system (2.1.4) may be written

$$\sum_{j=0}^{n} a_{ij}\xi_j = b_i, \qquad i = 0, \ldots, n, \tag{2.1.5}$$

where

$$a_{ij} = \int_{-1}^{1} x^{i+j} w(x)\, dx \tag{2.1.6}$$

and

$$b_i = \int_{-1}^{1} x^i f(x) w(x)\, dx. \tag{2.1.7}$$

In principle, then, we can determine $\xi_0, \ldots, \xi_n$ from (2.1.5), (which are called the "normal" equations) and thereby obtain q_n^* explicitly. However, when n is at all large, say $n \geq 7$, there appear, in the simple case that $w(x) \equiv 1$, formidable *numerical* difficulties in solving the normal equations. (For a discussion of the reason for these difficulties, see Forsythe [1].) It is possible to avoid these computational difficulties and find q_n^* in an extremely simple manner by observing that $\{1, x, x^2, \ldots, x^n\}$ is not the only set of functions that spans P_n, the space of polynomials of degree at most n. That is to say, if $p_0, p_1, \ldots, p_n \in P_n$ are linearly independent, then every $p \in P_n$ has a unique expression of the form

$$p = \alpha_0 p_0 + \alpha_1 p_1 + \cdots + \alpha_n p_n. \tag{2.1.8}$$

We are going to determine a set $\{p_0, p_1, \ldots, p_n\}$ which will be *orthogonal* with respect to the given weight function, $w(x)$. That is, we seek $p_0, p_1, \ldots, p_n \in P_n$ such that

$$\int_{-1}^{1} p_j(x) p_k(x) w(x)\, dx = 0, \qquad j \neq k, \qquad j, k = 0, \ldots, n. \tag{2.1.9}$$

If, in addition to (2.1.9), we also have

$$\int_{-1}^{1} p_j^2(x) w(x)\, dx = 1, \qquad j = 0, \ldots, n, \tag{2.1.10}$$

then $\{p_0, p_1, \ldots, p_n\}$ is called a set of *orthonormal* polynomials with respect to $w(x)$. Before proceeding to construct such an orthonormal set, let us see how it simplifies the least-squares approximation problem. Suppose $p_0, \ldots, p_n \in P_n$ satisfy (2.1.9) and (2.1.10). Then $p_0, p_1, \ldots, p_n$ are linearly independent (cf. Exercise 2.1). Let

$$q_n^*(x) = \lambda_0 p_0(x) + \lambda_1 p_1(x) + \cdots + \lambda_n p_n(x). \tag{2.1.11}$$

If we put $p = p_i$, $i = 0, \ldots, n$ successively in (2.1.2), we obtain

$$\int_{-1}^{1} p_i(x)q_n^*(x)w(x)\, dx = \int_{-1}^{1} p_i(x)f(x)w(x)\, dx, \qquad i = 0, \ldots, n$$

or, in view of (2.1.11),

$$\sum_{j=0}^{n} \lambda_j \int_{-1}^{1} p_i(x)p_j(x)w(x)\, dx = \int_{-1}^{1} p_i(x)f(x)w(x)\, dx, \qquad i = 0, \ldots, n.$$

$$(2.1.12)$$

But now, because of (2.1.9) and (2.1.10), Equation (2.1.12) becomes

$$\lambda_i = \int_{-1}^{1} p_i(x)f(x)w(x)\, dx, \qquad i = 0, \ldots, n. \qquad (2.1.13)$$

The normal equations are completely uncoupled and hence "self-solving" (or the matrix (a_{ij}) of (2.1.5) is now replaced by a diagonal matrix). Thus, the numerical difficulties in solving the normal equations disappear when we choose an orthogonal basis for P_n.

How can we obtain a set of polynomials orthonormal with respect to $w(x)$? We begin with some helpful notation. Let us put

$$\int_{-1}^{1} f(x)g(x)w(x)\, dx = (f, g). \qquad (2.1.14)$$

We call (f, g) the "inner product" of f and g.

First we define

$$\tilde{p}_0(x) = 1, \qquad \tilde{p}_1(x) = x - \frac{(1, x)}{(1, 1)}. \qquad (2.1.15)$$

Note that $(1, 1) > 0$ and $(\tilde{p}_0, \tilde{p}_1) = 0$. We wish to define an orthogonal set of polynomials $\{\tilde{p}_0, \tilde{p}_1, \ldots, \tilde{p}_k\}$ by mathematical induction. Suppose $\{\tilde{p}_0, \ldots, \tilde{p}_k\}$ form an orthogonal set with $\tilde{p}_i \in P_i$, $\tilde{p}_i \neq 0$, $i = 0, \ldots, k$; we wish to determine α_k and β_k so that

$$\tilde{p}_{k+1}(x) = (x - \alpha_k)\tilde{p}_k(x) - \beta_k\tilde{p}_{k-1}(x) \qquad (2.1.16)$$

is orthogonal to $\tilde{p}_1, \ldots, \tilde{p}_k$. (2.1.16) implies that

$$(\tilde{p}_{k+1}, \tilde{p}_i) = (x\tilde{p}_k, \tilde{p}_i) - \alpha_k(\tilde{p}_k, \tilde{p}_i) - \beta_k(\tilde{p}_{k-1}, \tilde{p}_i).$$

If $i < k - 1$, the inductive hypothesis implies that $(\tilde{p}_{k-1}, \tilde{p}_i) = (\tilde{p}_k, \tilde{p}_i) = 0$. Also $(x\tilde{p}_k, \tilde{p}_i) = (\tilde{p}_k, x\tilde{p}_i)$. Since $x\tilde{p}_i \in P_{k-1}$, we can write $x\tilde{p}_i = \gamma_0\tilde{p}_0 + \cdots + \gamma_{k-1}\tilde{p}_{k-1}$, from which we conclude that $(\tilde{p}_k, x\tilde{p}_i) = 0$, and hence

$$(\tilde{p}_{k+1}, \tilde{p}_i) = 0, \qquad i = 0, \ldots, k - 2.$$

$(\tilde{p}_{k+1}, \tilde{p}_k) = (x\tilde{p}_k, \tilde{p}_k) - \alpha_k(\tilde{p}_k, \tilde{p}_k)$. Let us choose

$$\alpha_k = \frac{(x\tilde{p}_k, \tilde{p}_k)}{(\tilde{p}_k, \tilde{p}_k)}; \qquad (2.1.17)$$

then

$$(\tilde{p}_{k+1}, \tilde{p}_k) = 0.$$

$(\tilde{p}_{k+1}, \tilde{p}_{k-1}) = (x\tilde{p}_k, \tilde{p}_{k-1}) - \beta_k(\tilde{p}_{k-1}, \tilde{p}_{k-1})$. Let us choose

$$\beta_k = \frac{(x\tilde{p}_k, \tilde{p}_{k-1})}{(\tilde{p}_{k-1}, \tilde{p}_{k-1})}; \qquad (2.1.18)$$

then

$$(\tilde{p}_{k+1}, \tilde{p}_{k-1}) = 0,$$

and since $\tilde{p}_{k+1} \in P_{k+1}$, $\{\tilde{p}_0, \ldots, \tilde{p}_{k+1}\}$ is an orthogonal set of the required kind. By mathematical induction, we conclude that $\{\tilde{p}_0, \tilde{p}_1, \ldots, \tilde{p}_n\}$, $\tilde{p}_i \in P_i$ is a set of orthogonal polynomials obtained from

$$\tilde{p}_0 = 1, \qquad \tilde{p}_1 = (x - \alpha_0)$$

by the three-term recurrence formula (2.1.16), for $k = 1, \ldots, n - 1$, where α_k is defined in (2.1.17) for $k = 0, \ldots, n - 1$ and β_k in (2.1.18) for $k = 1, \ldots, n - 1$.

We now define

$$p_i = \frac{\tilde{p}_i}{\|\tilde{p}_i\|_2}, \qquad i = 0, \ldots, n. \qquad (2.1.19)$$

The set $\{p_0, p_1, \ldots, p_n\}$ is orthonormal with respect to $w(x)$. Also, $p_i \in P_i$, and the leading coefficient of p_i is *positive* for $i = 0, \ldots, n$. There is no other orthonormal set of polynomials having these properties (cf. Exercise 2.3).

The least-squares approximation of degree n of f is, therefore, given by (2.1.11), where p_i is defined in (2.1.19) and

$$\lambda_i = (f, p_i), \qquad i = 0, \ldots, n. \qquad (2.1.20)$$

The use of orthogonal polynomials to obtain q_n^* has an added attraction. In order to obtain q_{n+1}^*, we need only find p_{n+1} by means of (2.1.16) and (2.1.19) and then determine λ_{n+1} by (2.1.20). The result is

$$q_{n+1}^* = q_n^* + \lambda_{n+1}p_{n+1}.$$

This is in marked contrast to best uniform approximation, where, in general, all of the coefficients of p_{n+1}^* are independent of the coefficients of p_n^*.

2.2 The Jacobi Polynomials

To carry our discussion further, we consider certain specific weight functions.

$$w(x) = (1 - x)^\alpha (1 + x)^\beta$$

is a weight function on I if $\alpha, \beta > -1$. An associated set of orthogonal polynomials is $\{P_j^{(\alpha,\beta)}(x)\}$, called the *Jacobi polynomials*. These polynomials are customarily standardized by

$$P_j^{(\alpha,\beta)}(1) = \frac{(\alpha + 1)(\alpha + 2)\cdots(\alpha + j)}{j!}. \tag{2.2.1}$$

There is a vast and important literature on the general theory of orthogonal polynomials and on the properties of the Jacobi polynomials. The reader who is interested should consult Szegö [1] or The Bateman Project [1]. We shall restrict our attention here to two of the most widely used sets of orthogonal polynomials, both instances of Jacobi polynomials.

Perhaps the simplest case is that where $w(x) \equiv 1$; that is, all points of I are equally weighted. The resulting polynomials are Jacobi polynomials ($\alpha = \beta = 0$) and are called the *Legendre polynomials*. Instead of denoting them by $P_j^{(0,0)}(x)$, we write simply $P_j(x)$. According to (2.2.1), the Legendre polynomials are standardized by the requirement that

$$P_n(1) = 1, \qquad n = 0, 1, 2, \ldots. \tag{2.2.2}$$

The recurrence relation (2.1.16) yields a set of orthogonal polynomials, call them $\{\tilde{P}_n\}$, which are standardized by having leading coefficient 1. Since $\tilde{P}_k$ is even for even k and odd for odd k (cf. Exercise 2.9), we see that $\alpha_k = 0$ in (2.1.16). Let us determine $\tilde{P}_k(1)$ to find the relationship between P_k and $\tilde{P}_k$. We claim that

$$\tilde{P}_k(1) = \frac{2^k}{\dbinom{2k}{k}}, \tag{2.2.3}$$

and will prove this by mathematical induction.

(2.2.3) is true for $k = 0, 1$ [$\binom{0}{0}$ is taken to be 1] by (2.1.15). Suppose that (2.2.3) holds for $k < j$ ($j \geq 2$); then

$$\tilde{P}_j(1) = \tilde{P}_{j-1}(1) - \frac{(\tilde{P}_{j-1}, \tilde{P}_{j-1})}{(\tilde{P}_{j-2}, \tilde{P}_{j-2})} \tilde{P}_{j-2}(1), \tag{2.2.4}$$

in view of (2.1.16) and Exercise 2.8. The right-hand side of (2.2.4) consists of known quantities by the inductive hypothesis and Exercise 2.10, and the result follows by some easy algebraic manipulation.

Thus

$$P_n = \binom{2n}{n} 2^{-n} \tilde{P}_n, \qquad n = 0, 1, 2, \ldots. \tag{2.2.5}$$

Therefore, the Legendre polynomials satisfy the recurrence relationship

$$P_{n+1}(x) = \binom{2n+2}{n+1} 2^{-(n+1)} x \tilde{P}_n(x) - \binom{2n+2}{n+1} 2^{-(n+1)} \frac{n^2}{4n^2 - 1} \tilde{P}_{n-1}(x)$$

$$= \frac{2n+1}{n+1} x P_n(x) - \frac{n}{n+1} P_{n-1}(x)$$

or

$$(n+1)P_{n+1}(x) = (2n+1)xP_n(x) - nP_{n-1}(x). \tag{2.2.6}$$

The first few Legendre polynomials are thus seen to be

$$P_0(x) = 1, \qquad P_1(x) = x, \qquad P_2(x) = \tfrac{3}{2}x^2 - \tfrac{1}{2}, \qquad P_3(x) = \tfrac{5}{2}x^3 - \tfrac{3}{2}x,$$
$$P_4(x) = \tfrac{35}{8}x^4 - \tfrac{15}{4}x^2 + \tfrac{3}{8}.$$

They are depicted in Figure 2.1.

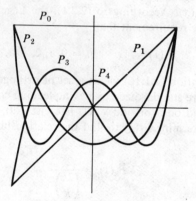

FIGURE 2.1

Next let us consider the Jacobi polynomials in the case $\alpha = \beta = -\tfrac{1}{2}$, so that

$$w(x) = (1 - x^2)^{-1/2}. \tag{2.2.7}$$

We saw [Exercise 1.16(c)] that the Chebyshev polynomials defined as

$$T_n(x) = \cos n\theta, \qquad n = 0, 1, 2, \ldots, \tag{2.2.8}$$

where $x = \cos \theta$, $0 \le \theta \le \pi$, are orthogonal with respect to the weight function given in (2.2.7). Thus, in view of (2.2.1),

$$P_j^{(-1/2, -1/2)} = \frac{1 \cdot 3 \cdot 5 \cdots (2j-1)}{2^j j!} T_j(x).$$

The recurrence relationship for the Chebyshev polynomials is

$$T_n(x) = 2xT_{n-1}(x) - T_{n-2}(x), \qquad n \geq 2, \qquad (2.2.9)$$

as we saw in Exercise 1.16(b), with $T_0(x) = 1$ and $T_1(x) = x$.

Since

$$T_n'(x) = \left(\frac{d}{d\theta} \cos n\theta\right) \left(-\frac{1}{\sin \theta}\right) = n \frac{\sin n\theta}{\sin \theta},$$

we conclude that

$$U_n(x) = \frac{\sin (n+1)\theta,}{\sin \theta} \qquad n = 0, 1, \ldots, \qquad (2.2.10)$$

where $x = \cos \theta$ is a polynomial of degree at most n. It is easy to verify that the polynomials (2.2.10) are orthogonal with respect to $w(x) = (1 - x^2)^{1/2}$ and hence are constant multiples of $P_n^{(1/2,1/2)}(x)$. Indeed, since, as L'Hôpital's rule reveals,

$$U_n(1) = n + 1,$$

we have

$$P_n^{(1/2,1/2)} = \frac{1 \cdot 3 \cdots (2n+1)}{2^n(n+1)!} U_n.$$

The polynomials $U_n(x)$ are called the Chebyshev polynomials of the second kind, and they will reappear in a later chapter. There are many identities connecting the polynomials (2.2.9) and (2.2.10) which are simply paraphrases of trigonometric identities. Some of them are given as exercises at the end of this chapter.

2.3 Approximation on a Finite Set of Points

We turn next to the least-squares approximation problem on a finite point set. Let X_m be a set of m distinct points, $x_1 < x_2 < \cdots < x_m$, of the interval $[-1, 1]$ and suppose the positive weight w_i is attached to the point x_i. Given f defined on X_m, then $q_n^* = q_n^*(X_m) \in P_n$ is called the least-square approximation f of degree n on X_m if

$$\sum_{i=1}^{m} [f(x_i) - q_n^*(x_i)]^2 w_i < \sum_{i=1}^{m} [f(x_i) - p(x_i)]^2 w_i$$

for any $p \in P_n$, $p \neq q_n^*(X_m)$. As we saw in the introductory chapter, $q_n^*(X_m)$ exists and is unique.

The discussion of the case of approximation on an interval can now be repeated with integrals everywhere replaced by summation over X_m. For example, the reader can easily verify the analogue of Theorem 2.1; namely, q_n^* is a best least-squares approximation to f on X_m if and only if

$$\sum_{i=1}^{m} [f(x_i) - q_n^*(x_i)]p(x_i)w_i = 0 \qquad (2.3.1)$$

for every $p \in P_n$. If we introduce the notation of the inner product of f and g, now defined by

$$(f, g) = \sum_{i=1}^{m} f(x_i)g(x_i)w_i, \qquad (2.3.2)$$

a repetition of the argument of Section 2.1 leads us to a set of orthogonal polynomials on X_m, defined by (2.1.15), (2.1.16), (2.1.17), and (2.1.18). However, one way the case of a finite point set differs from the continuous case is that any polynomial $p \in P_k$, $k \geq m$ containing

$$\omega(x) = (x - x_1)\cdots(x - x_m)$$

as a factor satisfies

$$(\omega, h) = \sum_{i=1}^{m} \omega(x_i)h(x_i)w(x_i) = 0$$

for *every* h, whereas only the zero polynomial has this property in the continuous case. This has the consequence that, in the case of a finite point set consisting of m points, the orthogonal polynomial $\tilde{p}_m$ defined by (2.1.16) must be $\omega(x)$. Hence, α_m is not defined in (2.1.17), and the recurrence relationship (2.1.16) breaks off at $k = m - 1$. (It is clear, however, that $\tilde{p}_{m+s}$ can be chosen to be any $p \in P_{m+s}$ having ω as a factor.)

As an example, let us consider the case where the points X_{m+1} are equally spaced in $[-1, 1]$; that is, we put

$$X_{m+1} = \left\{-1 + \frac{2(j-1)}{m}, \quad j = 1, \ldots, m+1\right\}. \qquad (2.3.3)$$

We suppose also that $w_j = 1$, $j = 1, \ldots, m + 1$. The resulting orthogonal polynomials $\tilde{g}_0, \ldots, \tilde{g}_{m+1}$ are called the *Gram polynomials*. The choice of points (2.3.3) and weights all equal to 1 implies that $\alpha_k = 0$ in (2.1.17), and so the Gram polynomials are even functions for even n and odd functions for odd n. For example, we have $\tilde{g}_0 = 1$, $\tilde{g}_1 = x$, $\tilde{g}_2 = x^2 - (m + 2)/(3m)$. We are now in a position to verify our statement on p. 42 by determining q_2, the least-squares approximation of degree 2 to $|x|$ on $X_5 = \{-1, -\frac{1}{2}, 0, \frac{1}{2}, 1\}$. We have that $g_0 = 5^{-1/2}$, $g_1 = (\frac{2}{5})^{1/2}x$, and $g_2 = (\frac{8}{7})^{1/2}(x^2 - \frac{1}{2})$ are orthonormal on X_5, and we have

$$\lambda_0 = (|x|, g_0) = 3(5)^{-1/2}, \quad \lambda_1 = (|x|, g_1) = 0, \quad \lambda_2 = (|x|, g_2) = (\tfrac{8}{7})^{1/2}(\tfrac{3}{4}).$$

Hence,

$$q_2 = \lambda_0 g_0 + \lambda_1 g_1 + \lambda_2 g_2 = \tfrac{6}{7}x^2 + \tfrac{6}{35}.$$

Suppose next that the set X_m consists of the zeros of the Chebyshev polynomial $T_m(x)$, so that

$$x_j = \cos(2j - 1)\frac{\pi}{2m}, \qquad j = 1, \ldots, m.$$

The reader should verify that the Chebyshev polynomials $T_0, T_1, \ldots, T_m$ are orthogonal on X_m with respect to the weights $w_j = 1, j = 1, \ldots, m$.

Exercise 1.16(d) says that, if X_{m+1} consists of the extrema of $T_m(x)$ (the points of I at which $|T_m(x)| = 1$), so that

$$x_j = \cos\frac{(j - 1)\pi}{m}, \qquad j = 1, \ldots, m + 1,$$

the Chebyshev polynomials $T_0, T_1, \ldots, T_m$ are orthogonal on X_{m+1} with respect to the weights $w_1 = w_{m+1} = \tfrac{1}{2}$, $w_j = 1, j = 2, \ldots, m$. Note, however, that although (2.1.16) yields

$$\tilde{p}_j = \tilde{T}_j, \qquad j = 0, \ldots, m,$$

$$\tilde{p}_{m+1} = (x^2 - 1)\tilde{U}_{m-1}.$$

2.4 Effectiveness as a Uniform Approximation

As a final observation, let us obtain some information about the effectiveness of a least-squares approximation as a uniform approximation. Such information is useful in view of the ease of obtaining least-squares approximations compared with best uniform approximations.

We shall treat the case in which the weight function is

$$w(x) = (1 - x^2)^{-1/2}, \tag{2.4.1}$$

so that the orthogonal polynomials are the Chebyshev polynomials. We choose this instance not only because of the widespread practical use of these particular least-squares approximations but also because the least-squares approximations of a function with respect to the weight (2.4.1) can be interpreted as the partial sums of the usual Fourier series of the function. Indeed, suppose $f(x)$ is continuous on $[-1, 1]$. Then $g(\theta) = f(\cos\theta)$ is continuous on $0 \le \theta \le \pi$ and, if we put $g(-\theta) = g(\theta)$, it is continuous on $-\pi \le \theta \le \pi$. Then the partial sums of its Fourier series are

$$s_n(g; \theta) = \frac{a_0}{2} + \sum_{k=1}^{n}(a_k \cos k\theta + b_k \sin k\theta),$$

(see p. 17). Since g is an even function, $b_k = 0$, $k = 1, \ldots, n$, and

$$s_n(g; \theta) = \frac{a_0}{2} + \sum_{k=1}^{n} a_k \cos k\theta = \frac{a_0}{2} \sum_{k=1}^{n} a_k T_k(x). \qquad (2.4.2)$$

The definition of the a_k [(1.1.17), p. 17] reveals that (2.4.2) is the least-squares approximation of degree n with respect to (2.4.1).

Now, according to Lemma 1.4, if $g(\theta)$ is any continuous function on $-\pi \le \theta \le \pi$ of period 2π,

$$s_n(g; \theta) = \frac{1}{\pi} \int_{-\pi}^{\pi} g(\phi + \theta)v_n(\phi)\, d\phi,$$

where

$$v_n(\phi) = \frac{1}{2} + \sum_{k=1}^{n} \cos k\phi.$$

The reader should have no difficulty in verifying that

$$v_n(\phi) = \frac{\sin\{(2n + 1)/2\}\phi}{2 \sin (\phi/2)}, \qquad (\phi \ne 2m\pi).$$

Thus,

$$s_n(g; \theta) = \frac{1}{2\pi} \int_{-\pi}^{\pi} g(\phi + \theta) \frac{\sin\{(2n + 1)/2\}\phi}{\sin (\phi/2)}\, d\phi$$

$$= \frac{1}{2\pi} \int_{0}^{\pi} [g(\theta + \phi) + g(\theta - \phi)] \frac{\sin\{(2n + 1)/2\}\phi}{\sin (\phi/2)}\, d\phi.$$

We have therefore established

LEMMA 2.1.   If $g(\theta)$ is continuous on $-\pi \le \theta \le \pi$, has period 2π, and satisfies

$$|g(\theta)| \le M, \qquad -\pi \le \theta \le \pi,$$

then its Fourier partial sums satisfy

$$\max_{-\pi \le \theta \le \pi} |s_n(g; \theta)| \le ML_n, \qquad (2.4.3)$$

where

$$L_n = \frac{1}{\pi} \int_{0}^{\pi} \frac{|\sin (n + \frac{1}{2})\phi|}{\sin (\phi/2)}\, d\phi. \qquad (2.4.4)$$

The numbers L_n are known as *Lebesgue's constants*.

Remark.   The (even) function

$$g(\theta) = \operatorname{sgn} \frac{\sin (n + \frac{1}{2})\theta}{\sin (\theta/2)} \qquad (2.4.5)$$

satisfies $s_n(g; 0) = L_n$ in view of (2.4.4). Now, $g(\theta)$ has a finite number of discontinuities (jumps) in $[-\pi, \pi]$ but, given any $\varepsilon > 0$ with a slight "smoothing" at the discontinuities, we obtain an (even) *continuous* function, $h(\theta)$, of period 2π, satisfying $|h(\theta)| \le 1$, $-\pi \le \theta \le \pi$, and

$$s_n(h; 0) > L_n - \varepsilon. \tag{2.4.6}$$

Thus, the bound in (2.4.3) is actually a least upper bound among all continuous g satisfying $|g| \le M$.

LEMMA 2.2

$$L_n = \frac{1}{\pi} \int_0^\pi \frac{|\sin (n + \frac{1}{2})\phi|}{\sin (\phi/2)} d\phi < 3 + \frac{4}{\pi^2} \log n. \tag{2.4.7}$$

Proof. If $n = 1$, the result is established by direct integration. Suppose that $n \ge 2$.

$$L_n = \frac{1}{\pi} \int_0^\pi \left| \frac{\sin n\phi}{\tan (\phi/2)} + \cos n\phi \right| d\phi \le \frac{1}{\pi} \int_0^\pi \frac{|\sin n\phi|}{\tan (\phi/2)} d\phi + \frac{1}{\pi} \int_0^\pi |\cos n\phi| \, d\phi.$$

Now

$$\int_0^\pi |\cos n\phi| \, d\phi = \frac{1}{n} \int_0^{n\pi} |\cos \phi| \, d\phi.$$

Let

$$I_n = \int_0^{n\pi} |\cos \phi| \, d\phi;$$

then

$$I_{k+1} - I_k = \int_{k\pi}^{(k+1)\pi} |\cos \phi| \, d\phi = \int_0^\pi |\cos (\phi + k\pi)| d\phi = \int_0^\pi |\cos \phi| \, d\phi = I_1 = 2$$

and, thus, $I_n = 2n$ and

$$\frac{1}{\pi} \int_0^\pi |\cos n\phi| \, d\phi = \frac{2}{\pi}.$$

Furthermore, $h(x) = \tan x - x$ satisfies $h(0) = 0$, $h'(x) = \sec^2 x - 1 \ge 0$. Hence, $\tan x \ge x$ for $0 \le x \le \pi/2$ and

$$\int_0^\pi \frac{|\sin n\phi|}{\tan (\phi/2)} d\phi \le 2 \int_0^\pi \frac{|\sin n\phi|}{\phi} d\phi.$$

Now, since

$$\int_0^\pi \frac{|\sin n\phi|}{\phi} d\phi = \int_0^{n\pi} \frac{|\sin \phi|}{\phi} d\phi = \sum_{k=0}^{n-1} \int_{k\pi}^{(k+1)\pi} \frac{|\sin \phi|}{\phi} d\phi$$

$$= \sum_{k=0}^{n-1} \int_0^\pi \frac{|\sin (\phi + k\pi)|}{\phi + k\pi} d\phi = \int_0^\pi \sin \phi \left\{ \sum_{k=0}^{n-1} \frac{1}{\phi + k\pi} \right\} d\phi,$$

we have

$$\int_0^\pi \frac{|\sin n\phi|}{\phi} \, d\phi = \int_0^\pi \frac{\sin \phi}{\phi} \, d\phi + \int_0^\pi \sin \phi \left[\sum_{k=1}^{n-1} \frac{1}{\phi + k\pi} \right] d\phi. \qquad (2.4.8)$$

It is known that the constant

$$\frac{2}{\pi} \int_0^\pi \frac{\sin \phi}{\phi} \, d\phi$$

is approximately 1.179 (cf. Zygmund [1, p. 61]). The integrand of the second integral on the right-hand side of (2.4.8) has nonnegative factors, and we obtain an upper bound for the second factor, namely, in $0 \le \phi \le \pi$,

$$\sum_{k=1}^{n-1} \frac{1}{\phi + k\pi} \le \frac{1}{\pi} \sum_{k=1}^{n-1} \frac{1}{k}.$$

But

$$\sum_{k=1}^{n-1} \frac{1}{k} < 1 + \int_1^{n-1} \frac{1}{x} \, dx = 1 + \log (n - 1)$$

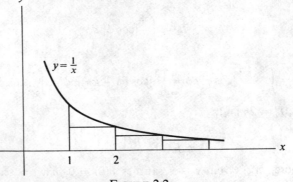

FIGURE 2.2

as Figure 2.2 shows. Therefore,

$$\int_0^\pi \sin \phi \left[\sum_{k=1}^{n-1} \frac{1}{\phi + k\pi} \right] d\phi < \frac{2}{\pi} \{1 + \log (n - 1)\}$$

and hence

$$L_n < \frac{2}{\pi} \int_0^\pi \frac{\sin \phi}{\phi} + \frac{4}{\pi^2} \{1 + \log (n - 1)\} + \frac{2}{\pi} < 3 + \frac{4}{\pi^2} \log n. \quad \blacksquare$$

THEOREM 2.2.   *Suppose that* $f \in C[-1, 1]$ *and* $S_n(f) = \|f - q_n^*\|$, *where* q_n^* *is the least-squares approximation with respect to* (2.4.1), *(that is,* $q_n^* = s_n(g; \theta)$, *where* $g(\theta) = f(\cos \theta)$, $g(-\theta) = g(\theta)$). *Then*

$$S_n(f) < \left(4 + \frac{4}{\pi^2} \log n\right) E_n(f).$$

Proof.   Suppose that p_n^* is the best uniform approximation to f on $[-1, 1]$, so that

$$|f(x) - p_n^*(x)| \le E_n, \qquad -1 \le x \le 1. \tag{2.4.9}$$

Let $f(\cos \theta) - p_n^*(\cos \theta)$ play the role of $g(\theta)$ in Lemma 2.1; then Lemmas 2.1, 2.2 imply

$$
\begin{aligned}
|s_n(f(\cos \theta) &- p_n^*(\cos \theta); \theta)| \\
&= |s_n(f(\cos \theta); \theta) - s_n(p_n^*(\cos \theta); \theta)| \\
&= |s_n(f(\cos \theta); \theta) - p_n^*(\cos \theta)| = |q_n^*(\cos \theta) - p_n^*(\cos \theta)| \\
&< \left(3 + \frac{4}{\pi^2} \log n\right) E_n. \tag{2.4.10}
\end{aligned}
$$

(2.4.10) and (2.4.9) prove the theorem. ∎

Exercises

2.1   Prove that if $\{p_0, \ldots, p_n\}$ is a set of orthogonal polynomials with respect to some weight function $w(x)$, then $p_0, p_1, \ldots, p_n$ are linearly independent if no $p_i = 0$.

[*Hint:* If $\alpha_0 p_0 + \alpha_1 p_1 + \cdots + \alpha_n p_n = 0$, multiply both sides by $p_i w$ and integrate over $[-1, 1]$ for each i.]

2.2   Suppose that

$$S_n(f; w) = \int_{-1}^{1} [f(x) - q_n^*(x)]^2 w(x)\, dx.$$

Show that

$$S_n(f; w) = \int_{-1}^{1} f^2(x) w(x)\, dx - \sum_{j=0}^{n} \lambda_j^2,$$

where the λ_j are as defined in (2.1.13).

2.3   Show (a) the set $\{p_0, p_1, \ldots, p_n\}$ of orthonormal polynomials (with respect to a given weight function) with $p_i \in P_i$ and leading coefficient of p_i positive for $i = 0, \ldots, n$ is unique. (b) the set $\{p_0, p_1, \ldots, p_n\}$ of *orthogonal* polynomials, with $p_i \in P_i$ having nonzero leading coefficient and $p_i(1)$ prescribed (but $\neq 0$) for $i = 0, \ldots, n$, is unique.

[*Hint:* Suppose that $\{q_0, q_1, \ldots, q_n\}$ is another such. Choose α_i so that $p_i - \alpha_i q_i \in P_{i-1}$. Then $\|p_i - \alpha_i q_i\|_2 = 0$.]

2.4   Prove that $f - q_n^*$ changes sign at $n + 1$ points of $(-1, 1)$ if $f \notin P_n$. (See p. 42 for an application of this fact.)

[*Hint:* If $f - q_n^*$ changes sign at at most $r \le n$ points of $(-1, 1)$, there exists $p \in P_n$ such that $(f - q_n^*)p \ge 0$ throughout I. Show that this contradicts (2.1.2) since $(f - q_n^*)p \in C(I)$ and $w(x) > 0$ except for a finite number of points.]

2.5   Show that the least-squares approximation of degree $n - 1$ to $f(x) = x^n$ is $q_{n-1}^* = x^n - \tilde{p}_n$, where $\tilde{p}_n$ is the orthogonal polynomial of degree n determined by $w(x)$ and (2.1.16).

2.6   Show that the orthogonal polynomial $\tilde{p}_n$ (and hence p_n) has n distinct simple zeros in $(-1, 1)$.

[*Hint:* Use Exercises 2.4 and 2.5.]

2.7   With the notation of Exercise 2.2, show that

$$\lim_{n \to \infty} S_n(f; w) = 0.$$

[*Hint:* $S_n(f; w) \le \|f - p_n^*\|_2^2$, where p_n^* is the best *uniform* approximation to f on I.]

2.8   Show that another form of (2.1.18) is

$$\beta_k = \frac{(\tilde{p}_k, \tilde{p}_k)}{(\tilde{p}_{k-1}, \tilde{p}_{k-1})}. \tag{2.4.11}$$

[*Hint:* Replace k by $k - 1$ in (2.1.16), then multiply both sides by $\tilde{p}_k w$ and integrate.]

2.9   Prove that the Jacobi polynomial $P_j^{(\alpha,\alpha)}$ is an odd function for odd j and an even function for even j.

[*Hint:* Use mathematical induction, (2.1.16) and the fact that $w(x)$ is an even function in this case.]

2.10   Show that, if $w(x) \equiv 1$, the orthogonal polynomials defined by (2.1.16) satisfy

$$(\tilde{p}_n, \tilde{p}_n) = \frac{2}{2n + 1} \tilde{p}_n^2(1), \qquad n = 0, 1, \ldots. \tag{2.4.12}$$

[*Hint:* Integration by parts yields

$$(\tilde{p}_n, \tilde{p}_n) = \tilde{p}_n^2(1) + \tilde{p}_n^2(-1) - 2n(\tilde{p}_n, \tilde{p}_n).]$$

2.11   Show that

$$(P_n, P_n) = \frac{2}{2n + 1}, \qquad n = 0, 1, 2, \ldots.$$

2.12 Prove Rodrigues' formula:

$$P_n(x) = \frac{2^{-n}}{n!} D^n[(x^2 - 1)^n)].$$ (2.4.13)

[*Hint:* First show, by repeated integration by parts, that

$$\int_{-1}^{1} D^n[(x^2 - 1)^n]x^k \, dx = 0, \qquad k = 0, 1, \ldots, n - 1;$$

then find the leading coefficient of $D^n[(x^2 - 1)^n]$.]

2.13 Show that the Legendre polynomials have the explicit representation

$$P_n(x) = 2^{-n} \sum_{j=0}^{[n/2]} (-1)^j \binom{n}{j} \binom{2n - 2j}{n} x^{n-2j},$$ (2.4.14)

and then check against the listing given on p. 54.

[*Hint:* Use Rodrigues' formula and the binomial expansion for $(x^2 - 1)^n$.]

2.14 Prove that

$$P_n'(x) = xP_{n-1}'(x) + nP_{n-1}(x).$$ (2.4.15)

[*Hint:*

$$D^{n+1}[(x^2 - 1)^n] = D^n\{D[(x^2 - 1)^{n-1}(x^2 - 1)]\} = 2nD^n[x(x^2 - 1)^{n-1}]$$
$$= 2n\{xD^n[(x^2 - 1)^{n-1}] + nD^{n-1}[(x^2 - 1)^{n-1}]\}$$

and use Rodrigues' formula.]

2.15 Show that

$$P_{n+1}'(x) - P_{n-1}'(x) = (2n + 1)P_n(x).$$ (2.4.16)

[*Hint:* Use the recurrence relationship (2.2.6) and Exercise 2.14.]

2.16 Show that

$$xP_n'(x) - P_{n-1}'(x) = nP_n(x).$$ (2.4.17)

[*Hint:* Use Exercises 2.14 and 2.15.]

2.17 Show that

$$(x^2 - 1)P_n'(x) = nxP_n(x) - nP_{n-1}(x).$$

[*Hint:* Multiply through in (2.4.17) by x and subtract (2.4.15).]

2.18 Show that

$$\frac{1 - x^2}{n^2}[P_n'(x)]^2 + P_n^2(x) = \frac{1 - x^2}{n^2}[P_{n-1}'(x)]^2 + P_{n-1}^2(x).$$ (2.4.18)

[*Hint:* Square and add (2.4.15) to the square of (2.4.17).]

2.19   Show that

$$\frac{1 - x^2}{n^2} [P_n'(x)]^2 + P_n^2(x) \le 1, \qquad (n \ge 1, \qquad |x| \le 1).$$

[*Hint:* Replacing n^2 by $(n - 1)^2$ on the right-hand side of (2.4.18) makes it larger.]

2.20   Prove that

$$|P_n(x)| \le 1, \qquad |x| \le 1.$$

2.21   Show that

$$T_n(x) = U_n(x) - xU_{n-1}(x) \quad \text{and} \quad (1 - x^2)U_{n-1}(x) = xT_n(x) - T_{n+1}(x).$$

2.22   Show that the Chebyshev polynomials of the second kind satisfy the three-term recurrence relationship

$$U_n(x) = 2xU_{n-1}(x) - U_{n-2}(x), \qquad n \ge 2 \qquad (2.4.19)$$

with $U_0(x) = 1$ and $U_1(x) = 2x$. [Compare with (2.2.9)!]

2.23   Show that, when X_{m+1} is defined as in (2.3.3) (equally spaced points) and all $w_j = 1$, the resulting orthogonal polynomials $\tilde{g}_0(m), \ldots, \tilde{g}_{m+1}(m)$ satisfy, for given k,

$$\lim_{m \to \infty} \tilde{g}_k(m) = \tilde{P}_k, \qquad (2.4.20)$$

where P_k is the Legendre polynomial normalized so that its leading coefficient is 1. Verify directly that (2.4.20) holds for $k = 2$.

The relationship between the least-squares approximation on a finite point set and the least-squares approximation on an interval as the finite point set "fills" the interval is discussed in Rice [1], where a result analogous to our Theorem 1.16 is proved.

2.24   Suppose that $f \in C[-\pi, \pi]$ and has period 2π; show that, if

$$\omega \left(f; [-\pi, \pi]; \frac{1}{n} \right) \log n \to 0, \qquad \text{as} \qquad n \to \infty,$$

then the Fourier series of f converges uniformly to f on $[-\pi, \pi]$.

2.25   Verify that, on $-1 \le x \le 1$,

$$|x| = \frac{2}{\pi} + \frac{4}{\pi} \sum_{k=1}^{\infty} \frac{(-1)^{k+1}}{4k^2 - 1} T_{2k}(x),$$

the convergence being uniform.

2.26   Verify that, for $f(x) = |x|$ we have, in the notation of Theorem 2.2,

$$S_n(f) = \begin{cases} \dfrac{2}{\pi n}, & n \text{ odd}, \\[3mm] \dfrac{2}{\pi(n + 1)}, & n \text{ even}; \end{cases}$$

hence, in view of Theorem 2.2, we have

$$E_n(|x|; [-1, 1]) > \frac{K}{n \log n},$$ (2.4.21)

where K is a constant.

Indeed, slightly more refined methods (see de La Vallée Poussin [1, p. 36]) show that the factor $\log n$ in the denominator of the right-hand side of (2.4.21) can be deleted.

  2.27   Show that

$$L_n > \frac{4}{\pi^2} \log n.$$

[*Hint:* Since $\sin (\phi/2) \le (\phi/2)$, it follows that

$$\frac{\pi}{2} L_n > \int_0^{(n + (1/2))\pi} \frac{|\sin \phi|}{\phi} \, d\phi.$$

A lower bound for this integral is found using the same methods that produced the upper bound on pp. 59, 60.]

CHAPTER 3

LEAST-FIRST-POWER APPROXIMATION

This chapter is devoted to the study of polynomials that minimize the integral of the absolute deviation from a given function. After characterizing best approximations and proving uniqueness, some special cases in which the best approximation is easily attainable are discussed. There follows consideration of approximation on a finite point set, and the chapter closes with a brief examination of some computational aspects of the problem.

3.1 Approximation on an Interval

In view of Theorem I.1, we know that, given $f(x)$ continuous on I: $[-1, 1]$, there exists a polynomial $r_n^* \in P_n$ such that

$$\|f - r_n^*\|_1 = \int_{-1}^{1} |f(x) - r_n^*(x)| \, dx \leq \|f - p\|_1 \qquad (3.1.1)$$

for all $p \in P_n$. In contrast to the case of least-squares approximation, we cannot conclude that r_n^* is unique from Theorem I.3, and must therefore leave aside, for the moment, the question of uniqueness. Any r_n^* satisfying (3.1.1) we call a least-first-power approximation to f out of P_n.

As is our custom, we shall begin by characterizing r_n^*. To this end we need some new notation. If $g \in C(I)$, let

$$Z(g) = \{x \in I / g(x) = 0\}; \qquad (3.1.2)$$

that is, $Z(g)$ is the set of zeros of g. Clearly, $Z(g)$ is a closed set. If $x \in Z(g)$ is such that every open interval of the real line that contains x also contains points that are *not* in $Z(g)$, then x is a boundary point of $Z(g)$.

DEFINITION. We call the boundary points of $Z(g)$ *essential zeros* of g, and denote them by $Z'(g)$.

DEFINITION.

$$\operatorname{sgn} g(x) = \begin{cases} 1 & \text{if } g(x) > 0, \\ 0 & \text{if } g(x) = 0, \\ -1 & \text{if } g(x) < 0. \end{cases} \tag{3.1.3}$$

The signum, or sign of $g(x)$ [sgn $g(x)$] is seen to be discontinuous in $(-1, 1)$ at precisely the essential zeros of g in $(-1, 1)$. Note that, if $Z'(g)$ consists of a finite number of points, then $Z(g)$ contains only a finite number of isolated points and a finite number of disjoint closed subintervals of I, for the end points of such subintervals are in $Z'(g)$. In this case, we denote the set of disjoint closed subintervals of I contained in $Z(g)$ by $Z_I(g)$, and have $Z(g) = Z_I(g) \cup Z'(g)$.

We can now characterize least-first-power approximations, as follows.

THEOREM 3.1.[1]   *Given $f \in C(I)$ and $r_n^* \in P_n$ such that $f - r_n^*$ has a finite number of essential zeros in I then r_n^* is a least-first-power approximation to f if and only if,*

$$\left| \int_{-1}^{1} \operatorname{sgn} [f(x) - r_n^*(x)] \cdot p(x)\, dx \right| \leq \int_{Z_I(f - r_n^*)} |p(x)|\, dx, \tag{3.1.4}$$

for every $p \in P_n$.

Remark.   If $Z_I(f - r_n^*)$ is empty, the integral on the right in (3.1.4) is taken to be zero, and the absolute value on the left in (3.1.4) is redundant.

Proof.   (i) Suppose that (3.1.4) holds. Then, for any $p \in P_n$,

$$\int_{-1}^{1} |f(x) - r_n^*(x)|\, dx$$

$$= \int_{-1}^{1} [f(x) - r_n^*(x)] \operatorname{sgn} [f(x) - r_n^*(x)]\, dx$$

$$= \int_{-1}^{1} [f(x) - p(x)] \operatorname{sgn} [f(x) - r_n^*(x)]\, dx$$

$$\qquad + \int_{-1}^{1} [p(x) - r_n^*(x)] \operatorname{sgn} [f(x) - r_n^*(x)]\, dx$$

$$\leq \int_{I - Z_I[f(x) - r_n^*(x)]} |f(x) - p(x)|\, dx$$

$$\qquad + \int_{Z_I[f(x) - r_n^*(x)]} |p(x) - r_n^*(x)|\, dx$$

$$= \int_{-1}^{1} |f(x) - p(x)|\, dx,$$

where we have used (3.1.4) (with $p - r_n^*$ playing the role of p) and the fact that $f(x) = r_n^*(x)$ on $Z_l(f(x) - r_n^*(x))$.

(ii) Suppose that (3.1.4) does not hold for some $p \in P_n$; then by replacing p by $-p$, if necessary, we have

$$\int_{-1}^{1} p(x) \operatorname{sgn} [f(x) - r_n^*(x)] \, dx > \int_{Z_l(f - r_n^*)} |p| \, dx. \qquad (3.1.5)$$

Let $x_1, \ldots, x_k$ be the points of $Z'(f - r_n^*)$ other than end points of intervals in $Z_l(f - r_n^*)$. Choose $\varepsilon > 0$ so small that, if $Z_i = [a_i, b_i]$, $i = 1, \ldots, m$ are the elements of $Z_l(f - r_n^*)$, the open intervals† $B_i = (a_i - \varepsilon, b_i + \varepsilon)$, $i = 1, \ldots, m$, $B_{m+i} = (x_i - \varepsilon, x_i + \varepsilon)$, $i = 1, \ldots, k$ are disjoint, and if

$$B = \bigcup_{i=1}^{m+k} B_i, \qquad A = I - B,$$

we have

$$\int_A p(x) \operatorname{sgn} [f(x) - r_n^*(x)] \, dx > \int_B |p| \, dx. \qquad (3.1.6)$$

A is a finite union of disjoint closed intervals; hence, it is closed. Since

$$A \cap Z(f - r_n^*) = \varnothing,$$

we know that there exists $m > 0$ such that

$$|f(x) - r_n^*(x)| \geq m, \qquad x \in A.$$

Since p is not the zero polynomial (why?), we can choose $\delta > 0$ so small that

$$0 < \delta \max_{x \in I} |p(x)| < m,$$

and, consequently,

$$\operatorname{sgn} [f(x) - r_n^*(x)] = \operatorname{sgn} [f(x) - r_n^*(x) - \delta p(x)], \qquad x \in A. \quad (3.1.7)$$

Our proof will now be completed by showing that, if we put $p_0(x) = r_n^*(x) + \delta p(x)$, then $\|f - p_0\|_1 < \|f - r_n^*\|_1$.

Now

$$\|f - p_0\|_1 = \int_B |f - p_0| \, dx + \int_A |f - p_0| \, dx$$

$$= \int_B |f - p_0| \, dx + \int_A (f - p_0) \operatorname{sgn} (f - p_0) \, dx.$$

---

† If some a_i, b_i, or x_i is ± 1, the corresponding B_i is half-open.

Hence, in view of (3.1.7) and (3.1.6),

$$\|f - p_0\|_1 = \int_B |f - p_0| \, dx + \int_A (f - p_0) \, \text{sgn} \, (f - r_n^*) \, dx$$

$$= \int_B |f - p_0| \, dx + \int_A (f - r_n^*) \, \text{sgn} \, (f - r_n^*) \, dx$$

$$- \delta \int_A p \, \text{sgn} \, (f - r_n^*) \, dx$$

$$< \int_B |f - p_0| \, dx + \int_A |f - r_n^*| \, dx - \int_B |\delta p| \, dx.$$

Now $A = I - B$, and so

$$\|f - p_0\|_1 < \|f - r_n^*\|_1 + \int_B (|f - p_0| - |f - r_n^*|) \, dx - \int_B |\delta p| \, dx.$$

But $|f - p_0| - |f - r_n^*| \le |\delta p|$, from which we conclude that

$$\|f - p_0\|_1 < \|f - r_n^*\|_1.$$

Since $p_0 \in P_n$, we have shown that r_n^* is not a least-first-power approximation. ∎

CORCLLARY 3.1.1.   *If $f \in C(I)$ and $r_n^* \in P_n$ are such that $f - r_n^*$ has a finite number of zeros in I, then r_n^* is a least-first-power approximation to f if and only if*

$$\int_{-1}^{1} \text{sgn} \, (f(x) - r_n^*(x)) \cdot p(x) \, dx = 0 \tag{3.1.8}$$

for all $p \in P_n$.

Notice the resemblance between (3.1.8) and (2.1.2).
We can now dispose of the uniqueness question.

THEOREM 3.2 (D. JACKSON).   *A continuous function, $f(x)$, on I has a unique least-first-power approximation out of P_n.*

Proof.   Suppose that $p_1, p_2 \in P_n$ are least-first-power approximations to f, $p_1 \ne p_2$ and

$$\|f - p_1\|_1 = \|f - p_2\|_1 = m. \tag{3.1.9}$$

Since the set of best approximations is convex according to Theorem I.2, we conclude that

$$p_0 = \frac{p_1 + p_2}{2}$$

is also a least-first-power approximation to f; that is,

$$\|f - p_0\|_1 = m. \tag{3.1.10}$$

(3.1.9) and (3.1.10) imply that

$$0 = \int_{-1}^{1} \left\{ |f(x) - p_0(x)| - \frac{1}{2} |f(x) - p_1(x)| - \frac{1}{2} |f(x) - p_2(x)| \right\} dx.$$

$$(3.1.11)$$

Since

$$|f(x) - p_0(x)| = \left| f(x) - \frac{p_1(x) + p_2(x)}{2} \right|$$

$$= \frac{1}{2} |[f(x) - p_1(x)] + [f(x) - p_2(x)]|$$

$$\leq \frac{1}{2} |f(x) - p_1(x)| + \frac{1}{2} |f(x) - p_2(x)|,$$

we learn that the integrand in (3.1.11) is nonpositive. It is also continuous and therefore must be identically zero. This means that, if $f - p_0$ has k distinct zeros in I. $k \leq n$. For suppose that $k > n$, and let the zeros of $f - p_0$ be $x_1, \ldots, x_k$. Then, for $i = 1, \ldots, k$,

$$0 = |f(x_i) - p_0(x_i)| - \frac{1}{2}|f(x_i) - p_1(x_i)| - \frac{1}{2}|f(x_i) - p_2(x_i)|$$
$$= -\frac{1}{2}|f(x_i) - p_1(x_i)| - \frac{1}{2}|f(x_i) - p_2(x_i)|,$$

from which we conclude that

$$f(x_i) - p_1(x_i) = f(x_i) - p_2(x_i) = 0.$$

Therefore, $p_1 - p_2 \in P_n$ has $k > n$ zeros, and so $p_1 = p_2$, which is contrary to our assumption.

Corollary 3.1.1 is now applicable with $r_n^* = p_0$ and tells us that

$$\int_{-1}^{1} \text{sgn} \, [f(x) - p_0(x)] \cdot p(x) \, dx = 0 \qquad (3.1.12)$$

for all $p \in P_n$. Let $t_1 < t_2 < \cdots < t_s$ be those zeros of $f - p_0$ located in $(-1, 1)$ at which $f - p_0$ changes sign. Then, certainly, $s \leq k \leq n$, and if we put $t_0 = -1$ and $t_{s+1} = 1$, (3.1.12) may be rewritten (possibly after being multiplied by -1) as

$$\sum_{j=0}^{s} (-1)^{s-j} \int_{t_j}^{t_{j+1}} p(x) \, dx = 0. \qquad (3.1.13)$$

If we put $p(x) = (x - t_1) \cdots (x - t_s)$, then $p \in P_n$ and

$$\text{sgn} \int_{t_j}^{t_{j+1}} p(x) \, dx = (-1)^{s-j}, \qquad j = 0, \ldots, s,$$

contradicting (3.1.13). The assumption that $p_1 \neq p_2$ has led to a contradiction. ∎

It is rather remarkable that, for a certain subset of $C(I)$, the least-first-power approximation can be determined explicitly quite simply. If $f \notin P_n$ and $f \in C(I)$, we shall say that f is *adjoined* to P_n if $f - p$ has at most $n + 1$ distinct zeros in I for every $p \in P_n$. For example, any $p_{n+1} \in P_{n+1}$ (but $p_{n+1} \notin P_n$) is adjoined to P_n. Also, any function having a continuous nonvanishing $(n + 1)$st derivative on I is adjoined to P_n (see Exercise 3.2).

THEOREM 3.3. *Let g be adjoined to P_n, and let p_0 be its least-first-power approximation out of P_n. Then $g - p_0$ changes sign at exactly $n + 1$ distinct points, $x_1, \ldots, x_{n+1}$, in $(-1, 1)$. Suppose that $f \in C(I)$ and r_n^* is its least-first-power approximation. Then, if $f - r_n^*$ changes sign at most $n + 1$ times in $(-1, 1)$ and has a finite number of distinct zeros in I, $f - r_n^*$ changes sign exactly at the points $x_1, \ldots, x_{n+1}$. (This is true, in particular if f is adjoined to P_n.)*

Proof. $g - p_0$ has at most $n + 1$ distinct zeros in $(-1, 1)$ since g is adjoined to P_n. But (cf. Exercise 3.1) $g - p_0$ changes sign at least $n + 1$ times in $(-1, 1)$ hence, exactly $n + 1$ times. Call the points at which $g - p_0$ changes sign $x_1, \ldots, x_{n+1}$. Similarly, $f - r_n^*$ changes sign at exactly $n + 1$ distinct points, $y_1, \ldots, y_{n+1}$, of $(-1, 1)$.

Let p_1 be the unique polynomial in P_n that agrees with g at $y_1, \ldots, y_{n+1}$. Since g is adjoined to P_n, $g - p_1$ has no zeros in I other than $y_1, \ldots, y_{n+1}$. We claim that $g - p_1$ changes sign at y_i, $i = 1 \ldots n + 1$. Suppose that $g - p_1$ does not change sign at y_j. Indeed, let us assume that $g(x) - p_1(x) > 0$ in $0 < |x - y_j| \leq \delta$, where δ is less than the distance from y_j to any other zero of $g - p_1$. (The case $g - p_1 < 0$ may be treated similarly.) Given $\varepsilon > 0$, let $p_2 \in P_n$ be defined by

$$p_2(x) = p_1(x) + \varepsilon \prod_{i=1, i \neq j}^{n+1} \frac{x - y_i}{y_j - y_i}.$$

Suppose that $0 < h < \delta$ and

$$\left| \prod_{i=1, i \neq j}^{n+1} \frac{x - y_i}{y_j - y_i} \right| \leq M, \qquad x \in I.$$

Let $\min\{[g(y_j + h) - p_1(y_j + h)], [g(y_j - h) - p_1(y_j - h)]\} = A$. Then $A > 0$. Choose $\varepsilon < A/M$; then

$$g(y_j) - p_2(y_j) = g(y_j) - p_1(y_j) - \varepsilon = -\varepsilon < 0,$$

while

$$g(y_j \pm h) - p_2(y_j \pm h) = g(y_j \pm h) - p_1(y_j \pm h) + p_1(y_j \pm h) - p_2(y_j \pm h)$$

$$= A - \varepsilon \prod_{i=1, i \neq j}^{n+1} \frac{y_j \pm h - y_i}{y_j - y_i}.$$

Since

$$\prod_{i=1, i \neq j}^{n+1} \frac{y_j \pm h - y_i}{y_j - y_i} > 0$$

for $0 \leq h < \delta$, we conclude that $g(y_j \pm h) - p_2(y_j \pm h) > A - \varepsilon M > 0$, and, hence, that $g - p_2$ has two zeros in $(y_j - \delta, y_j + \delta)$ in addition to the n zeros y_i, $i = 1, \ldots, n + 1$, $i \neq j$, contradicting the assumption that g is adjoined to P_n.

Thus, throughout I, either sgn $(g - p_1) = $ sgn $(f - r_n^*)$ or sgn $(g - p_1) = -$ sgn $(f - r_n^*)$. But in either case, since r_n^* is the least-first-power approximation to f, Theorem 3.1 implies that

$$\int_{-1}^{1} \text{sgn} \, [f(x) - r_n^*(x)] \cdot p(x) \, dx = 0$$

for all $p \in P_n$, and so

$$\int_{-1}^{1} \text{sgn} \, [g(x) - p_1(x)] \cdot p(x) \, dx = 0$$

for all $p \in P_n$. Now, Corollary 3.1.1 implies that p_1 is the least-first-power approximation to g, and therefore, by Theorem 3.2, $p_1 = p_0$, and hence $\{x_1, \ldots, x_{n+1}\} = \{y_1, \ldots, y_{n+1}\}$. ∎

Theorem 3.3 will be particularly valuable if we can find $x_1, \ldots, x_{n+1}$ for some g adjoined to P_n, for then the least-first-power approximation to any f adjoined to P_n is simply the polynomial that takes on the value $f(x_i)$ at x_i, $i = 1, \ldots, n + 1$. The obvious adjoined function to consider is x^{n+1}. We have

THEOREM 3.4.   *If p_0 is the least-first-power approximation to x^{n+1} out of P_n on I, then*

$$x^{n+1} - p_0 = \tilde{U}_{n+1}, \tag{3.1.14}$$

where $\tilde{U}_{n+1}$ is the Chebyshev polynomial of the second kind normalized so that its leading coefficient is 1; that is (see p. 55)

$$\tilde{U}_{n+1}(x) = \frac{1}{2^{n+1}} \frac{\sin (n + 2)\theta}{\sin \theta}, \qquad x = \cos \theta.$$

Hence, $x^{n+1} - p_0$ changes sign at

$$x_j = \cos \frac{j\pi}{n + 2}, \qquad j = 1, \ldots, n + 1. \tag{3.1.15}$$

Proof.   Define p_0 to be $x^{n+1} - \tilde{U}_{n+1}$. All will be shown if we can verify that

$$\int_{-1}^{1} [\text{sgn} \, \tilde{U}_{n+1}] \cdot p(x) \, dx = 0 \tag{3.1.16}$$

for all $p \in P_n$, according to Corollary 3.1.1. Let us write

$$p(x) = a_0 U_0(x) + a_1 U_1(x) + \cdots + a_n U_n(x).$$

Then, if we put $x = \cos \theta$, (3.1.16) is seen to be equivalent to

$$\int_0^\pi \text{sgn} \, [(\sin (n + 2)\theta][a_0 \sin \theta + a_1 \sin 2\theta + \cdots + a_n \sin (n + 1)\theta] \, d\theta = 0.$$

$$(3.1.17)$$

Now

$$I = \int_0^\pi \text{sgn} \, [\sin (n + 2)\theta] \cdot \sin k\theta \, d\theta = \frac{1}{2} \int_{-\pi}^\pi \text{sgn} \, [\sin (n + 2)\theta] \sin k\theta \, d\theta$$

$$= \frac{1}{2} \, \text{Im} \, K,$$

where

$$K = \int_{-\pi}^\pi \text{sgn} \, [\sin (n + 2)\theta] e^{ik\theta} \, d\theta = \int_{-\pi + (\pi/(n+2))}^{\pi + (\pi/(n+2))} \text{sgn} \, [\sin (n + 2)\theta] e^{ik\theta} \, d\theta$$

$$= -e^{i(k\pi/(n+2))} K.$$

Thus, for $k = 1, \ldots, n + 1$, $K = 0$; hence $I = 0$ and (3.1.16) is verified. ∎

COROLLARY 3.4.1.   *If f is adjoined to P_n, its least-first-power approximation out of P_n is the unique $r_n^* \in P_n$, which satisfies*

$$r_n^* \left(\cos \frac{j\pi}{n + 2} \right) = f \left(\cos \frac{j\pi}{n + 2} \right), \qquad j = 1, \ldots, n + 1.$$

3.2   *Approximation on a Finite Set of Points*

Let X_m be a set of m points $x_1 < x_2 < \cdots < x_m$ of I, and suppose that the positive weight w_i is attached to the point x_i. Given f defined on X_m, $r_n^* = r_n^*(X_m) \in P_n$ is called a least-first-power approximation to f out of P_n on X_m if

$$\sum_{i=1}^m |f(x_i) - r_n^*(x_i)| w_i \le \sum_{i=1}^m |f(x_i) - p(x_i)| w_i \qquad (3.2.1)$$

for all $p \in P_n$. We shall adopt the notation

$$\|g\|_{X_m} = \sum_{i=1}^m |g(x_i)| w_i,$$

so that (3.2.1) may be written

$$\|f - r_n^*\|_{X_m} \le \|f - p\|_{X_m}.$$

The existence of a least-first-power approximation of given degree is assured by our general existence result in the introductory chapter. As we saw there, however, the uniqueness question in this case is open. And, indeed, our first result will be to show, by an example, that there need not be a unique least-first-power approximation on a finite point set.

Let $X_2 = \{-1, 1\}$ and take $f(-1) = -1$, $f(1) = 1$, $w_1 = w_2 = 1$, and $n = 0$. Then, for any constant α, satisfying $-1 \leq \alpha \leq 1$, we have

$$\|f - \alpha\|_{X_2} = 2 < \|f - \beta\|_{X_2},$$

where $|\beta| > 1$.

The characterization of best approximations is the precise analogue of the case of the interval, but it is much simpler since no function can have more than m zeros on X_m.

THEOREM 3.5. r_n^* is a least-first-power approximation to f on X_m, if and only if

$$\left| \sum_{i=1}^{m} \text{sgn} \left[f(x_i) - r_n^*(x_i) \right] p(x_i) w_i \right| \leq \sum_{i \in Z(f - r_n^*)} |p(x_i)| w_i, \qquad (3.2.2)$$

for all $p \in P_n$, where, with a harmless abuse of usage, $Z(f - r_n^*)$ here denotes the set of indices, i, for which x_i is a zero $f - r_n^*$.

The proof is a considerably simplified repetition of the arguments used in proving Theorem 3.1, and we leave it to the reader. The proof yields the following corollary, which is of interest in the finite point set case.

COROLLARY 3.5.1.   If the inequality holds in (3.2.2) for all $p \in P_n$, $p \neq 0$ then r_n^* is unique.

We shall see later (Corollary 3.6.1) that the converse of this corollary is also true.

As an example of Theorem 3.5, let us find the least-first-power approximation to any f on X_m out of P_0 (the constants) in the case $w_i = 1$, $i = 1, \ldots, m$. Let us renumber the points $x_1, \ldots, x_m$, so that $f(x_1) \leq f(x_2) \leq \cdots \leq f(x_m)$. We choose

$$c = f(x_{(m+1)/2}), \qquad m \text{ odd}, \qquad (3.2.3)$$

$$f(x_{m/2}) \leq c \leq f(x_{(m/2)+1}), \qquad m \text{ even}, \qquad (3.2.4)$$

then $c = r_n^*$; that is, c is a least-first-power constant approximation to f on X_m. For proof, we invoke (3.2.2). [Note that (3.2.2) does *not* depend on the order of the points $x_1, \ldots, x_m$.] Suppose m odd and $f(x_r) = f(x_{r+1}) = \cdots = f(x_{(m+1)/2}) = \cdots = f(x_s)$, while $f(x_{r-1}) < f(x_r)$, $(r > 1)$, and $f(x_s) < f(x_{s+1})$, $(s < m)$. Then if p is any constant,

$$\left| \sum_{i=1}^{m} \text{sgn} \, [f(x_i) - c] \cdot p \right| = |p| \left| \sum_{i=1}^{r-1} (-1) + \sum_{i=s+1}^{m} 1 \right| = |p| \, |m - s - r + 1|,$$

while

$$\sum_{i=r}^{s} |p| = |p|(s - r + 1).$$

Since $s \geq (m + 1)/2$, $2s \geq m + 1$, and $s - r + 1 > m - s - r + 1$. Since $r \leq (m + 1)/2$, $2r \leq m + 1$, and $s - r + 1 > -(m - s - r + 1)$. Hence, $s - r + 1 > |m - s - r + 1|$, and we conclude from (3.2.2) that $r_0^* = c$ is the unique least-first-power constant approximation when m is odd. Note that c is the *median* of $f(x_1), \ldots, f(x_m)$.

If m is even, we consider two cases:

(a) $f(x_{(m/2)}) < f(x_{(m/2)+1})$.

Now, if c lies in the interval (3.2.4),

$$\left| \sum_{i=1}^{m} \text{sgn} \, [f(x_i) - c] \cdot p \right| = |p| \left| \sum_{i=1}^{(m/2)} (-1) + \sum_{i=(m/2)+1}^{m} 1 \right| = 0,$$

and so, according to (3.2.2), we have an interval (3.2.4) of best approximations.

(b) $f(x_{(m/2)}) = f(x_{(m/2)+1})$.

We now conclude, as in the case of odd m, that $r_0^* = c$, as defined in (3.2.4), is the unique best approximation.

We continue our study of approximation on a finite point set with a detailed examination of the properties of the set, B, of least-first-power approximations out of P_n to a given f on X_m. According to Theorem I.2 and Exercises I.10 and I.11, B is a compact convex set. We single out certain special polynomials in B, namely, the *extreme points* of B. These are points of B that are *not* midpoints of any segment in B. More precisely, we have the following

DEFINITION. If K is a convex set in a linear space, p is an *extreme point* of K if $p \in K$ and

$$p = \frac{p_1 + p_2}{2},$$

where $p_1, p_2 \in K$ implies that $p = p_1 = p_2$.

We shall see that the extreme points of B are of particular importance among the least-first-power approximations.

THEOREM 3.6. *If $p_0 \in B$ and for some $p \in P_n$, $p \neq 0$*

$$\| \text{sgn} \, (f - p_0) \cdot p \|_{X_m} = \left| \sum_{i=1}^{m} \text{sgn} \, [f(x_i) - p_0(x_i)] p(x_i) w_i \right| \qquad (3.2.5)$$

$$= \sum_{i \in Z(f - p_0)} |p(x_i)| w_i,$$

then there exist real numbers a, b, with $a < b$ such that $p_t = p_0 + tp \in B$ for $a \leq t \leq b$, while $p_t \notin B$ for $t > b$ or $t < a$.
If

$$\sum_{i \in Z(f - p_0)} |p(x_i)| w_i = 0, \qquad (3.2.6)$$

then $ab < 0$; otherwise, $ab = 0$.

Proof. Since

$$\min_{i \notin Z(f - p_0)} |f(x_i) - p_0(x_i)| > 0,$$

we observe that there exists $\varepsilon > 0$ such that, for $i \notin Z(f - p_0)$ and $|t| < \varepsilon$,

$$\text{sgn} \, [f(x_i) - p_0(x_i)] = \text{sgn} \, [f(x_i) - p_t(x_i)].$$

Therefore,

$$\| f - p_t \|_{X_m} - \| f - p_0 \|_{X_m}$$

$$= \sum_{i \notin Z(f - p_0)} \{ [f(x_i) - p_t(x_i)] - [f(x_i) - p_0(x_i)] \} \, \text{sgn} \, [f(x_i) - p_0(x_i)] w_i$$

$$+ \sum_{i \in Z(f - p_0)} |t p(x_i)| w_i$$

$$= -t \sum_{i \notin Z(f - p_0)} p(x_i) \, \text{sgn} \, [f(x_i) - p_0(x_i)] w_i + |t| \sum_{i \in Z(f - p_0)} |p(x_i)| w_i.$$

$$(3.2.7)$$

Case (i). If (3.2.6) holds, the right-hand side of (3.2.7) is zero and $p_t \in B$ for $|t| < \varepsilon$. We now choose a to be the least and b the greatest value of t for which p_t remains in B. Then $a \leq -\varepsilon$, $b \geq \varepsilon$, and $ab < 0$.

Case (ii). If

$$\sum_{i \in Z(f - p_0)} |p(x_i)| w_i > 0 \, ;$$

then, in view of (3.2.5), if $t \neq 0$, the right-hand side of (3.2.7) is 0 if t has the same sign as $\tau = \sum p(x_i) \, \text{sgn} \, [f(x_i) - p_1(x_i)] w_i$ and is positive otherwise. Thus, if $\tau > 0$, then $p_t \in B$ for $0 \leq t < \varepsilon$, and we choose $a = 0$ and let b be the largest value of t for which $p_t \in B$. If $\tau < 0$, then $p_t \in B$ for $-\varepsilon < t \leq 0$, and we choose $b = 0$ and let a be the smallest value of t for which p_t remains in B. ∎

COROLLARY 3.6.1. *If r_n^* is unique, the inequality holds in* (3.2.2) *for all* $p \in P_n, p \neq 0$.

COROLLARY 3.6.2. *If q is an extreme point of B, then $Z(f - q)$ is not empty.*

Proof. If $Z(f - q) = \varnothing$, then (3.2.2) implies that (3.2.5) and (3.2.6) hold with $p_0 = q$. But then, for $\varepsilon > 0$ and sufficiently small, $p_t \in B$ for $|t| \leq \varepsilon$, and $q = p_0 = \frac{1}{2}(p_\varepsilon + p_{-\varepsilon})$. Thus, q is not an extreme point of B. ▮

COROLLARY 3.6.3. *If q_1 and q_2 are distinct extreme points of B, then*

$$Z(f - q_1) \neq Z(f - q_2).$$

Proof. Suppose that $Z(f - q_1) = Z(f - q_2)$. Let $r = (q_1 + q_2)/2$. Then $r \in B$, and

$$\sum_{i \in Z(f - q_1)} |r(x_i) - q_1(x_i)| w_i > 0. \tag{3.2.8}$$

For if the sum in (3.2.8) were zero, then (3.2.5) would hold with $p_0 = q_1$ and $p = r - q_1$, as would (3.2.6), and the Theorem would imply that q_1 is not an extreme point of B. But (3.2.8) implies that there exists $i \in Z(f - q_1)$ such that $i \notin Z(r - q_1)$ on the one hand, while on the other hand it is clear from the definition of r that $Z(f - q_2) = Z(f - q_1) \subset Z(f - r)$. This contradiction establishes the corollary. ▮

THEOREM 3.7. *The set of extreme points of B is not empty and consists of a finite number of polynomials, $q_1, \ldots, q_s$. Moreover, B is the convex hull of $\{q_1, \ldots, q_s\}$; that is, if $q \in B$, then there exist $\lambda_1, \ldots, \lambda_s \geq 0$ satisfying*

$$\sum_{i=1}^{s} \lambda_i = 1,$$

such that

$$q = \sum_{i=1}^{s} \lambda_i q_i.$$

Proof. That B can have, at most, a finite number of extreme points follows from Corollaries 3.6.2 and 3.6.3 and the fact that X_m is a finite set of points. The remainder of the theorem is a consequence of (a weak version of) the Krein–Milman Theorem (the convex hull of $\{q_1, \ldots, q_s\}$ being closed), which says that, in a finite dimensional normed linear space, the closure of the convex hull of a compact set B equals the closure of the convex hull of its extreme points (cf. Valentine [1]). ▮

THEOREM 3.8. *Every extreme point of B agrees with f on at least $n + 1$ points of X_m.*

Proof. Suppose the theorem to be false; then there exists an extreme point

$p_0 \in B$, and $Z(f - p_0)$ consists of at most n points. Let $p \in P_n$, $p \neq 0$, be chosen to have $Z(f - p_0)$ as its set of zeros; then

$$\sum_{i \in Z(f - p_0)} |p(x_i)| w_i = 0,$$

so that (3.2.5) and (3.2.6) hold. Theorem 3.6 then implies that p_0 is not an extreme point of B. ∎

3.3   Some Computational Aspects

The problem of finding a least-first-power approximation out of P_n on X_m is equivalent to a linear programming problem, as we demonstrate next.

Let $p(x) = a_0 + a_1 x + \cdots + a_n x^n$. Our problem, then, is to determine $a_0, \ldots, a_n$ so that

$$\sum_{i=1}^{m} |f_i - (a_0 + a_1 x_i + \cdots + a_n x_i^n)| w_i \qquad (3.3.1)$$

is a minimum for given $f_1, \ldots, f_m$ and positive $w_1, \ldots, w_m$. Let us put

$$e_i = [f_i - (a_0 + a_1 x_i + \cdots + a_n x_i^n)] w_i, \qquad i = 1, \ldots, m,$$

and write

$$e_i = u_i - v_i, \qquad i = 1, \ldots, m,$$

where, if $e_i \geq 0$,

$$u_i = e_i, \qquad v_i = 0,$$

while, if $e_i \leq 0$,

$$u_i = 0, \qquad v_i = e_i.$$

Then

$$u_i \geq 0, \qquad v_i \geq 0$$

for $i = 1, \ldots, m$.

Let us also put

$$\beta = \min_{j = 0, \ldots, n} a_j$$

and

$$\alpha_{n+1} = \max (0, -\beta), \qquad \alpha_j = a_j + \alpha_{n+1}, \qquad j = 0, \ldots, n,$$

so that

$$\alpha_j \geq 0, \qquad j = 0, \ldots, n + 1.$$

Consider, now, the linear programming problem of minimizing

$$\sum_{i=1}^{m} (u_i + v_i) \qquad (3.3.2)$$

subject to the constraints

$$f_i w_i - \sum_{j=0}^{n} \alpha_j x_i^j w_i + \alpha_{n+1} \sum_{j=0}^{n} x_i^j w_i - u_i + v_i = 0, \qquad i = 1, \ldots, m, \qquad (3.3.3)$$

where the variables u_i, v_i, α_j, α_{n+1} are nonnegative. A solution for this problem must have the property that $u_i v_i = 0$, and hence that $|u_i - v_i| = u_i + v_i$, $i = 1, \ldots, m$. For if, for some $k = 1, \ldots, m$, $u_k v_k > 0$, then $u_k > 0$ and $v_k > 0$; hence, upon replacing u_k and v_k by $u_k - \min(u_k, v_k)$ and $v_k - \min(u_k, v_k)$, respectively, in (3.3.2) and (3.3.3), (3.3.2) is decreased while the left-hand side of (3.3.3) remains unaltered. Thus, the minimization of (3.3.2) subject to (3.3.3) is seen to be equivalent to the minimization of (3.3.1). The solution of (3.3.2), (3.3.3) can be accomplished by means of the simplex method (cf. p. 43; the interested reader should be sure to consult Barrodale and Young [1] and Usow [1]).†

Finally, let us examine the connection between the approximation problem on an interval and that on a discrete subset of the interval. Taken together with the immediately preceding discussion, this will provide us with a computational procedure for the continuous case.

Let $0 = t_0 < t_1 < \cdots < t_m = 1$. $\Delta t_i = t_i - t_{i-1}$ is the length of the interval $I_i : t_{i-1} \leq x \leq t_i$, $i = 1, \ldots, m$, and

$$\delta_m = \max_{i=1,\ldots,m} \Delta t_i.$$

Let X_m denote the set of points $x_1 < x_2 < \cdots < x_m$, where $x_i \in I_i$, $i = 1, \ldots, m$. If such a partitioning of $I : [0, 1]$ and choice of X_m is made for each positive integer m, subject only to the restriction that $\delta_m \to 0$ as $m \to \infty$, then, for each function g continuous on I, the definition of the (Riemann) integral is

$$\int_0^1 g(t)\, dt = \lim_{m \to \infty} \sum_{i=1}^m g(x_i)\, \Delta t_i.$$

Moreover,

LEMMA 3.1

$$\left| \int_0^1 g(t)\, dt - \sum_{i=1}^m g(x_i)\, \Delta t_i \right| \leq \omega(g; I, \delta_m) = \omega(\delta_m),$$

where the modulus of continuity, $\omega(\delta_m)$, is as defined on p. 14.

Proof

$$\left| \int_0^1 g(t)\, dt - \sum_{i=1}^m g(x_i)\, \Delta t_i \right| = \left| \sum_{i=1}^m \int_{t_{i-1}}^{t_i} [g(t) - g(x_i)]\, dt \right|$$

$$\leq \sum_{i=1}^m \int_{t_{i-1}}^{t_i} |g(t) - g(x_i)|\, dt \leq \omega(\delta_m) \sum_{i=1}^m \Delta t_i$$

$$\leq \omega(\delta_m).$$

† See, also, Rubio [1].

We now can state

THEOREM 3.9.[2] *Suppose that p^* is the least-first-power approximation out of P_n to a given continuous function, f, on I: $[0, 1]$. Let p_m be a least-first-power approximation out of P_n to f on X_m with weights $w_i = \Delta t_i$, $i = 1, \ldots, m$. If $\delta_m \to 0$, then p_m converges uniformly to p^* on I as $m \to \infty$.*

Proof. Let

$$\mu = \int_0^1 |f(t) - p^*(t)| \, dt; \qquad \mu_m = \sum_{j=1}^m |f(x_j) - p_m(x_j)| \, \Delta t_j.$$

In view of the definitions of p_m and of the integral, we have, for m sufficiently large,

$$\mu_m \leq \sum_{j=1}^m |f(x_j) - p^*(x_j)| \, \Delta t_j < \mu + 1;$$

hence,

$$\sum_{j=1}^m |p_m(x_j)| \, \Delta t_j - \sum_{j=1}^m |f(x_j)| \, \Delta t_j < \mu + 1.$$

$$\sum_{j=1}^m \Delta t_j = t_m - t_0 = 1$$

implies that

$$\sum_{j=1}^m |f(x_j)| \, \Delta t_j \leq \max_{j=1,\ldots,m} |f(x_j)| \leq \max_{t \in I} |f(t)| = M,$$

and so we obtain

$$\sum_{j=1}^m |p_m(x_j)| \, \Delta t_j \leq \mu + 1 + M = A_1. \qquad (3.3.4)$$

We show next that the polynomials $\{p_m\}$ are uniformly bounded on I. To this end, let $J_1, \ldots, J_{n+1}$ be mutually disjoint closed subintervals of I, each of length $d > 0$. Choose m sufficiently large so that the sum of the lengths of the intervals I_j that are contained in J_i is greater than $d/2$ for $i = 1, \ldots, n+1$. This is possible since $\delta_m \to 0$. Let the set of j for which $I_j \subset J_i$ be denoted by $\mathcal{J}_i$. Then, in view of (3.3.4), we have

$$\frac{d}{2} \min_{j \in \mathcal{J}_i} |p_m(x_j)| \leq \sum_{j \in \mathcal{J}_i} |p_m(x_j)| \, \Delta t_j \leq \sum_{j=1}^m |p_m(x_j)| \, \Delta t_j \leq A_1; \quad i = 1, \ldots, n+1.$$

Thus, in each interval J_i, $i = 1, \ldots, n+1$, there exists a point (of X_m), call it ξ_i, defined by

$$\min_{j \in \mathcal{J}_i} |p_m(x_j)| = |p_m(\xi_i)|$$

such that

$$|p_m(\xi_i)| \leq \frac{2A_1}{d} = A_2; \qquad i = 1, \ldots, n+1.$$

The points $\xi_1, \ldots, \xi_{n+1}$ depend on m, but A_2 is independent of m. By the Lagrange interpolation formula [cf. (1.3.5), p. 35, or the beginning of Chapter 4],

$$p_m(t) = \sum_{i=1}^{n+1} p_m(\xi_i) \left\{ \prod_{j=1, j \neq i}^{n+1} \frac{(t - \xi_j)}{(\xi_i - \xi_j)} \right\}.$$

If, anticipating the notation of Chapter 4, we write

$$l_i(t) = \prod_{j=1, j \neq i}^{n+1} \frac{(t - \xi_j)}{(\xi_i - \xi_j)},$$

and note that, since the J_i are closed and disjoint, $i = 1, \ldots, n + 1$, there exists a positive number c such that $|\xi_i - \xi_j| \geq c$, $i, j = 1, \ldots, n + 1$, $i \neq j$, then, for all $t \in I$,

$$|l_i(t)| \leq \left(\frac{2}{c}\right)^n, \qquad i = 1, \ldots, m + 1,$$

and thus, for every positive integer m,

$$|p_m(t)| \leq (n + 1)A_2 \left(\frac{2}{c}\right)^n = A_3,$$

where A_3 is independent of m.

If we put

$$p_m(t) = a_{0,m} + a_{1,m}t + \cdots + a_{n,m}t^n,$$

then, in view of the inequality of V. Markov [(1.2.22) on p. 32], there exist constants $B_0, \ldots, B_n$ depending only on n, such that

$$|a_{j,m}| = \frac{|p_m^{(j)}(0)|}{j!} \leq B_j, \qquad j = 0, \ldots, n; \qquad \text{all } m.$$

The sequence $\{a_{0,m}\}$ is bounded and hence has a convergent subsequence $\{a_{0,\sigma}\}$ with limit α_0. The sequence $\{a_{1,\sigma}\}$ is bounded and hence has a convergent subsequence $\{a_{1,\tau}\}$ with limit α_1. Continuing in this fashion, we extract convergent subsequences from each successively defined subsequence until finally we obtain a subsequence of the positive integers, $\{m_i\}$, with the property that, as $i \to \infty$,

$$a_{j,m_i} \to \alpha_j, \qquad j = 0, \ldots, n.$$

Thus, the sequence $\{p_{m_i}(t)\}$ converges uniformly on I to

$$p_0(t) = \alpha_0 + \alpha_1 t + \cdots + \alpha_n t^n.$$

For each i, $|p_{m_i}(t)| \leq A_3$, and so, in view of (1.2.22),

$$|p'_{m_i}(t)| \leq A_4, \qquad t \in I,$$

where A_4 depends only on n. Thus, as a consequence of the mean-value theorem,

$$|p_{m_i}(t') - p_{m_i}(t'')| = |p'_{m_i}(\eta)| \, |t' - t''| \le A_4 |t' - t''|$$

for any t', $t'' \in I$ and some η between t' and t''. Thus, for any $\delta > 0$, ($\delta < 1$), and all i,

$$\omega(p_{m_i}; I, \delta) \le A_4 \delta. \tag{3.3.5}$$

Since $\omega(g_1 + g_2; I; \delta) \le \omega(g_1; I; \delta) + \omega(g_2; I; \delta)$ and $\omega(|g|; I; \delta) \le \omega(g; I; \delta)$, we conclude from (3.3.5) and Lemma 3.1 that, for each i,

$$\left| \int_0^1 |f(t) - p_{m_i}(t)| \, dt - \sum_{j=1}^{m_i} |f(x_j) - p_{m_i}(x_j)| \, \Delta t_j \right| \le \omega([f - p_{m_i}]; I; \delta_{m_i})$$
$$\le \omega(f; I; \delta_{m_i}) + A_4 \delta_{m_i}. \tag{3.3.6}$$

If we put

$$\mu_0 = \int_0^1 |f(t) - p_0(t)| \, dt,$$

then, given $\varepsilon > 0$, there exists i_0 such that, for $i > i_0$, by the uniform convergence of $\{p_{m_i}\}$ to p_0,

$$\left| \mu_0 - \int_0^1 |f(t) - p_{m_i}(t)| \, dt \right| < \frac{\varepsilon}{2} \quad \text{and} \quad \left| \int_0^1 |f(t) - p_{m_i}(t)| \, dt - \mu_{m_i} \right| < \frac{\varepsilon}{2}$$

in view of (3.3.6). Thus, for $i > i_0$,

$$|\mu_0 - \mu_{m_i}| < \varepsilon, \tag{3.3.7}$$

and, in particular, recalling the definition of μ,

$$\mu \le \mu_0 < \mu_{m_i} + \varepsilon.$$

But, i_0 may also be chosen so large that, for $i > i_0$,

$$\left| \mu - \sum_{j=1}^{m_i} |f(x_j) - p^*(x_j)| \, \Delta t_j \right| \le \omega([f - p^*]; I; \delta_{m_i}) < \varepsilon,$$

and, by definition,

$$\mu_{m_i} \le \sum_{j=1}^{m_i} |f(x_j) - p^*(x_j)| \, \Delta t_j.$$

Hence,

$$\mu_{m_i} \le \mu + \varepsilon,$$

which, taken together with (3.3.7), yields $\mu_0 = \mu$. Thus, we have shown that p_0 is a least-first-power approximation to f on I out of P_n. By the uniqueness theorem (Theorem 3.2), $p_0 = p^*$. Since the sequence $\{p_m\}$ is now seen to have exactly one limit point, namely p^*, the theorem is proved. ∎

Remark. With a particular choice of the t_i, say, $t_i = i/m$ so that $\Delta t_i = 1/m$ and $\delta_m = 1/m$, it is not difficult to estimate the difference $\mu - \mu_m$ as a function of m, by estimating the relevant quantities as they appear in the proof.

Exercises

3.1 If r_n^* is a least-first-power approximation to f, $f - r_n^*$ has a finite number of essential zeros in I, and $Z_I(f - r_n^*)$ is empty, show that $f - r_n^*$ changes sign at least $n + 1$ times in $(-1, 1)$.

[*Hint:* Examine the argument at the end of the proof of Theorem 3.2.]

3.2 Show that, if f has a continuous nonvanishing $(n + 1)$st derivative on I, it is adjoined to P_n.

[*Hint:* Apply Rolle's Theorem $n + 1$ times.]

3.3 Show that

$$\int_{-1}^{1} |p(x)| \, dx > \int_{-1}^{1} |\tilde{U}_n(x)| \, dx = \frac{1}{2^{n-1}}$$

for all $p \in P_n$ with leading coefficient 1, $p \neq \tilde{U}_n$.

3.4 Suppose that

$$0 < m < f^{(n+1)}(x) < M, \qquad x \in I: [-1, 1].$$

Let

$$\hat{U}_k = \frac{\tilde{U}_k}{k!}.$$

Show that:

(i) $\phi(x) = f(x) - r_n^*(x)$

has precisely the same zeros as $\hat{U}_{n+1}$ in I, namely,

$$\eta_j = \cos \frac{j\pi}{n+2}, \qquad j = 1, \ldots, n+1. \tag{3.3.8}$$

and changes sign at each of its zeros, just as $\hat{U}_{n+1}$ does.

(ii) $m\hat{U}_{n+1}^{(n+1)}(x) < \phi^{(n+1)}(x) < M\hat{U}_{n+1}^{(n+1)}(x), \qquad x \in I,$

3.5 With the conditions of Exercise 3.4 holding, show that the zeros of $\phi - m\hat{U}_{n+1}$ and $M\hat{U}_{n+1} - \phi$ in I are exactly as in (3.3.8), and that each function changes sign at each of these zeros.

3.6   Show that, in the interval $[\cos \pi/(n + 2), 1]$,

$$0 < \phi - m\hat{U}_{n+1},$$
$$0 < M\hat{U}_{n+1} - \phi,$$
$$0 < \hat{U}_{n+1},$$
$$0 < \phi,$$

and, hence, that

$$m|\hat{U}_{n+1}| < |\phi| < M|\hat{U}_{n+1}|$$

throughout I.

3.7   Show that, if

$$0 < A \le |f^{(n+1)}(x)| \le B, \qquad x \in I,$$

then

$$\int_{-1}^{1} |f(x) - p(x)| \, dx \ge \frac{A}{(n + 1)!} 2^{-n}$$

for all $p \in P_n$, and

$$\int_{-1}^{1} |f(x) - r_n^*(x)| \, dx \le \frac{B}{(n + 1)!} 2^{-n}.$$

In particular, for example, if $p_0 \in P_n$ is the least-first-power approximation to e^x on $[-1, 1]$, then

$$\frac{e^{-1}}{2^n(n + 1)!} \le \int_{-1}^{1} |e^x - p_0| \, dx \le \frac{e}{2^n(n + 1)!}.$$

3.8   Show that, if $f^{(n)}(x)$ is continuous and does not vanish on I, then

$$\int_{-1}^{1} |f(x)| \, dx \ge \frac{2^{-n}}{(n + 1)!} \min_{x \in I} |f^{(n)}(x)|$$

with equality if, and only if, $f = cU_n$, c being any nonzero constant.

3.9   Find the least-first-power constant approximation to a given function on X_m in the case where w_i, $i = 1, \ldots, m$ are arbitrary positive weights. When is the result unique?

3.10   Show that the least-first-power approximation of the form ax (that is, by means of lines passing through the origin) to $f(x)$ on X_m with $w_i = 1$, $i = 1, \ldots, m$, is obtained as follows: Renumber the points $(x_i, f(x_i))$ for which $x_i \ne 0$ so that

$$\frac{f(x_1)}{x_1} \ge \frac{f(x_2)}{x_2} \ge \cdots \ge \frac{f(x_l)}{x_l} \qquad [l = m \quad \text{or} \quad m - 1].$$

Now let r be the smallest integer such that

$$2 \sum_{i=1}^{r} |x_i| \geq \sum_{i=1}^{l} |x_i|;$$

then

$$\frac{f(x_r)}{x_r} x$$

is a least-first-power approximation of the required form. Discuss uniqueness.

[*Hint:* While Theorem 3.5 cannot be applied, since we are not admitting all polynomials of degree 1 as approximators, show that an exact analogue of Theorem 3.5 holds when the family of approximators consists of $\{ax\}$. Then apply this result. What is the result when the w_i are arbitrary positive numbers?]

This problem has an interesting history. The result is the algebraic formulation by Laplace of a geometric method due to Boscovich. A fascinating account of Boscovich's approach and its subsequent ramifications as well as the history of the relative merits of least-first-power and least-squares fitting of data can be found in Eisenhart [1].

3.11   Given X_3, show that the least-first-power approximation by polynomials of degree at most 1, with $w_1 = w_2 = w_3 = 1$, to f on X_3 is provided by the line passing through $(x_1, f(x_1))$ and $(x_3, f(x_3))$.

Suppose $m = n + 2, q_1, \ldots, q_s$ to be the extreme points of B and $f \notin P_n$ in Exercises 3.12 through 3.15.

3.12   Show that there is exactly one point, call it $x_i \in X_m$, at which $q_i(x_i) \neq f(x_i)$, the points $x_1, \ldots, x_s$ are distinct, and each of $q_1, \ldots, q_s$ agrees with f on $x_{s+1}, \ldots, x_{n+2}$.

3.13   If

$$p_0 = \lambda_1 q_1 + \cdots + \lambda_s q_s,$$

where $\lambda_j > 0, j = 1, \ldots, s$, and $\lambda_1 + \cdots + \lambda_s = 1$, show that p_0 agrees with f precisely on $x_{s+1}, \ldots, x_{n+2}$.

3.14   If $p_0 \in B$ agrees with f on $n + 1$ points, show that p_0 is an extreme point of B.

Remark.   The extreme points of B are now seen to be found among the (at most) $n + 2$ polynomials that agree with f on some $n + 1$ points of X_m, and hence all of B can be determined by direct examination of (at most) $n + 2$ polynomials.

3.15   Suppose that $q \in B$ is an extreme point of B agreeing with f on $x_1, \ldots, x_{n+1}$. Show that, for any $g \notin P_n$, the unique $p_0 \in P_n$ that agrees with g

on $x_1, \ldots, x_{n+1}$ is a least-first-power approximation to g. (Compare with Theorem 3.3.)

[*Hint:* g can be represented uniquely as $g = p + af$ for some $p \in P_n$ and some constant a.]

Notes, Chapter 3

(1) A much more general result and a thorough discussion of the uniqueness question can be found in Kripke and Rivlin [1].

(2) This theorem is due to Motzkin and Walsh [1]. A more general result can be found in Kripke [1].

CHAPTER 4

POLYNOMIAL AND SPLINE INTERPOLATION

One of the most direct ways to approximate a function on an interval, or a finite set of points, is to attempt to obtain a polynomial, or another simple approximator, which takes the same values as the function at a certain number of points in the domain of the function. This procedure is called *interpolation*. We have already seen how to find interpolating polynomials (cf. p. 35). There is a vast literature on the various useful representations of an interpolating polynomial (cf. Steffenson [1], Hildebrand [1], Modern Computing Methods [1]). We shall not enter into that area but will restrict our attention to a study of the efficacy of interpolation as an approximating process. We shall first consider interpolation by polynomials and then *spline* interpolation, that is, interpolation by functions that are piecewise polynomials.

4.1 General Results

Given distinct $x_1, \ldots, x_n$ satisfying $-1 \le x_i \le 1$, $i = 1, \ldots, n$, and $f(x)$, a continuous function on $I: [-1, 1]$. Let us put

$$f_i = f(x_i), \qquad i = 1, \ldots, n.$$

The polynomials

$$l_j(x) = \frac{(x - x_1)(x - x_2) \cdots (x - x_{j-1})(x - x_{j+1}) \cdots (x - x_n)}{(x_j - x_1)(x_j - x_2) \cdots (x_j - x_{j-1})(x_j - x_{j+1}) \cdots (x_j - x_n)},$$

$$j = 1, \ldots, n \quad (4.1.1)$$

satisfy $l_j \in P_{n-1}$ and

$$l_j(x_i) = \begin{cases} 0, & j \ne i, \\ 1, & j = i, \end{cases} \qquad i, j = 1, \ldots, n. \qquad (4.1.2)$$

(In case $n = 1$, we take $l_1(x) \equiv 1$.) We call $l_1(x), \ldots, l_n(x)$ the *fundamental polynomials* for interpolation at $x_1, \ldots, x_n$.

$$L_{n-1}(x) = \sum_{i=1}^{n} f_i l_i(x), \qquad (4.1.3)$$

is a polynomial of degree at most $n - 1$, which has the value f_i at x_i, $i = 1, \ldots, n$. Moreover, if $p \in P_{n-1}$ and $p(x_i) = f_i$, $i = 1, \ldots, n$, then $L_{n-1} - p \in P_{n-1}$ has n distinct zeros, namely, $x_1, \ldots, x_n$, and hence $L_{n-1} = p$. Thus, L_{n-1}, as defined in (4.1.3), is the unique member of P_{n-1} that interpolates to f at $x_1, \ldots, x_n$. This unique interpolating polynomial, when written in the form (4.1.3), is called the *Lagrange interpolating polynomial* [to $f(x)$ at $x_1, \ldots, x_n$].

We wish to study the approximating power of interpolating polynomials as the number of points of interpolation or *nodes*, as they are called, increases. To this end, we consider an infinite triangular array of nodes,

$$X: \quad \begin{matrix} x_1^{(1)} & & & \\ x_1^{(2)} & x_2^{(2)} & & \\ x_1^{(3)} & x_2^{(3)} & x_3^{(3)} & \\ \vdots & & & \\ x_1^{(n)} & x_2^{(n)} & \cdots & x_n^{(n)} \\ \vdots & & & \end{matrix} \qquad (4.1.4)$$

where for $n = 1, 2, \ldots, -1 \le x_1^{(n)} < x_2^{(n)} < \cdots < x_n^{(n)} \le 1$. Given f, the rows of X determine a sequence of interpolating polynomials $L_0, L_1, \ldots, L_{n-1}$ being the unique polynomial in P_{n-1} that agrees with f at the nodes of the nth row of X. The notation $L_k(x)$ is shorthand for $L_k(f, X; x)$, in which the subscript indicates that we have an element of P_k that agrees with f at the nodes of the $(k + 1)$st row of X, evaluated at x. The polynomial $L_k(f, X; x)$ is defined by (4.1.3) in terms of fundamental polynomials $l_j^{(k)}$,† $j = 1, \ldots, k + 1$, which have the property exhibited in (4.1.2). Let

$$G_k = G_k(f; X) = \|f - L_k\|, \qquad k = 0, 1, \ldots,$$

the norm being the uniform norm on I. Then, recalling the notation of p. 11, we have the following comparison of G_k and $E_k(f; I)$.

THEOREM 4.1

$$G_k \le E_k \left(1 + \max_{-1 \le x \le 1} \sum_{j=1}^{k+1} |l_j^{(k)}(x)| \right), \qquad k = 0, 1, \ldots. \qquad (4.1.5)$$

† Here (k) is an index and should not be confused with the notation for the kth derivative.

Proof. Let $p^* \in P_k$ be the best (uniform) approximation to f on I. Then

$$|f(x) - L_k(x)| \le |f(x) - p^*(x)| + |p^*(x) - L_k(x)|. \tag{4.1.6}$$

By the uniqueness of the Lagrange interpolating polynomial,

$$p^*(x) = L_k(p^*, X; x),$$

and hence

$$p^*(x) - L_k(f, X; x) = L_k(p^*, X; x) - L_k(f, X; x) = L_k(p^* - f, X; x).$$

(4.1.6) now yields

$$|f(x) - L_k(x)| \le E_k + L_k(p^* - f, X; x). \tag{4.1.7}$$

But

$$|L_k(p^* - f, X; x)| \le \max_{-1 \le x \le 1} |p^*(x) - f(x)| \cdot \max_{-1 \le x \le 1} \sum_{j=1}^{k+1} |l_j^{(k)}(x)|. \tag{4.1.8}$$

The theorem now follows from (4.1.7) and (4.1.8). ∎

The function

$$\lambda_k(X; x) = \sum_{j=1}^{k+1} |l_j^{(k)}(x)|, \qquad k = 0, 1, \ldots, \tag{4.1.9}$$

which appears in (4.1.5), is called the *Lebesgue function* of order k of X. Note that it does not depend on f. The quantity

$$\Lambda_k(X) = \max_{-1 \le x \le 1} \lambda_k(X; x)$$

is called the *Lebesgue constant* of order k of X. (4.1.5) may now be written concisely as

$$G_k \le E_k(1 + \Lambda_k), \qquad k = 0, 1, \ldots. \tag{4.1.10}$$

But according to Jackson's Theorem (Theorem 1.4), $E_k \le 6\omega(1/k)$ and, hence,

$$G_k \le 6(1 + \Lambda_k)\omega(1/k). \tag{4.1.11}$$

Thus, insofar as Theorem 4.1 is informative, it tells us that, given X and $f \in C(I)$, the sequence of interpolating polynomials converges uniformly to f on I if $\Lambda_k \omega(1/k) \to 0$ as $k \to \infty$. If we know more about f, say that it has a certain number of derivatives, we can use the appropriate variant of Jackson's Theorem (see the discussion following Theorem 1.4) to bound E_k in (4.1.10) and obtain results analogous to (4.1.11). We must still estimate the size of $\Lambda_k(X)$ and study the implications of these estimates. This we do in the next section.

4.2 The Size of the Lebesgue Constants

Let X be a given triangular array of nodes in I, and suppose that $f(x) \in C(I)$. If we put $x = \cos\theta$, I corresponds to $0 \le \theta \le \pi$.

Let us fix our attention on the $(n + 1)$st row of X and let

$$x_i^{(n+1)} = \cos\theta_i, \qquad i = 1, \ldots, n + 1.$$

$g(\theta) = f(\cos\theta)$ is continuous on $[0, \pi]$, and if we put $g(-\theta) = g(\theta)$, it is continuous on $[-\pi, \pi]$ and has period 2π. With this change of variables we define

$$U_n(g; \theta) = L_n(f, X; x) = \sum_{i=1}^{n+1} g(\theta_i) \prod_{j=1, j \ne i}^{n+1} \frac{\cos\theta - \cos\theta_j}{\cos\theta_i - \cos\theta_j}. \qquad (4.2.1)$$

$U_n(g; \theta)$ is the unique cosine polynomial of degree at most n, which agrees with g at $\theta_1, \ldots, \theta_{n+1}$. Given any real number ϕ, we define a function of θ

$$g_\phi(\theta) = \frac{g(\theta + \phi) + g(\theta - \phi)}{2}$$

We then have

LEMMA 4.1 (FABER)

$$\frac{1}{\pi} \int_{-\pi}^{\pi} U_n(g_\phi; \theta)_\phi \, d\phi = s_n(g; \theta) + \frac{1}{2\pi} \int_{-\pi}^{\pi} g(t) dt \qquad (4.2.2)$$

where $s_n(g; \theta)$ is the partial sum of the Fourier series of g (cf. p. 17).

Proof.

$$\frac{1}{\pi} \int_{-\pi}^{\pi} U_n(g_\phi; \theta) d\phi = \frac{1}{\pi} \int_{-\pi}^{\pi} U_n(g_\phi - s_n(g_\phi); \theta)_\phi d\phi$$
$$+ \frac{1}{\pi} \int_{-\pi}^{\pi} U_n(s_n(g_\phi); \theta)_\phi d\phi = A_n + B_n.$$

As we saw on p. 58

$$s_n(g_\phi; \theta) = \frac{1}{2\pi} \int_{-\pi}^{\pi} g_\phi(\theta + t) \frac{\sin((2n+1)/2)t}{\sin(t/2)} \, dt = s_n(g; \theta)_\phi.$$

$$(4.2.3)$$

But since $U_n(s_n(g_\phi))$ is the *unique* cosine polynomial of degree at most n that agrees with $s_n(g_\phi)$ at $\theta_1, \ldots, \theta_{n+1}$ and $s_n(g_\phi)$ is a cosine polynomial of degree at most n, we have, in view of (4.2.3)

$$U_n(s_n(g_\phi); \theta)_\phi = s_n(g_\phi; \theta)_\phi = s_n(g; \theta)_{\phi\phi}$$

and hence

$$B_n = s_n(g;\theta) + \frac{1}{2\pi} \int_{-\pi}^{\pi} g(t)dt.$$

The proof of the lemma will now be completed by showing that $A_n = 0$.

$$U_n(g_\phi - s_n(g_\phi); \theta \pm \phi)$$

$$= \sum_{i=1}^{n+1} [g_\phi(\theta_i) - s_n(g_\phi; \theta_i)] \prod_{j=1, j \neq i}^{n+1} \frac{\cos(\theta \pm \phi) - \cos\theta_j}{\cos\theta_i - \cos\theta_j}. \qquad (4.2.4)$$

Since the Fourier coefficients of $g - s_n(g)$ with index $\leq n$ are all zero, we observe that

$$\int_{-\pi}^{\pi} [g(\theta_i \pm \phi) - s_n(g; \theta_i \pm \phi)]\cos k(\theta_i \pm \phi)d\phi = 0,$$

and

$$\int_{-\pi}^{\pi} [g(\theta_i \pm \phi) - s_n(g; \theta_i \pm \phi)]\sin k(\theta_i \pm \phi)d\phi = 0,$$

for $k = 0,1,...,n$. But for each $i = 1,...,n+1$,

$$\prod_{j=1, j \neq i}^{n+1} \frac{\cos(\theta \pm \phi) - \cos\theta_j}{\cos\theta_i - \cos\theta_j} = \prod_{j=1, j \neq i}^{n+1} \frac{\cos((\theta + \theta_i) - (\theta_i \pm \phi)) - \cos\theta_j}{\cos\theta_i - \cos\theta_j}$$

is a trigonometric polynomial in $\theta_i \pm \phi$ of order at most n. It therefore follows at once, from (4.2.4) that $A_n = 0$, and the lemma is proved.

It is clear, from (4.2.2) that, given X, n, and g, there exists a ϕ such that

$$|s_n(g;\theta)| \leq 2|U_n(g_\phi;\theta)_\phi| + \|g\|; \qquad (4.2.5)$$

for example, for each θ, pick the ϕ for which $(U_n)_\phi$ assumes its maximum absolute value. Let g be the (even) continuous function h described in the Remark following Lemma 2.1, satisfying $|h| \leq 1$ and

$$s_n(h;0) > L_n - 1.$$

Then (4.2.5) implies that

$$|U_n(h_\phi; \phi)| = |U_n(h_\phi; 0)_\phi| > \frac{L_n}{2} - 1.$$

But since $|h_\phi| \leq 1$,

$$|U_n(h_\phi; \phi)| \leq \max_{0 \leq \phi \leq \pi} \sum_{i=1}^{n+1} \prod_{j=1, j \neq i}^{n+1} \left| \frac{\cos \phi - \cos \theta_j}{\cos \theta_i - \cos \theta_j} \right| = \Lambda_n(X).$$

If we recall Exercise 2.27, we have proved

THEOREM 4.2

$$\Lambda_n(X) > \frac{2}{\pi^2} \log n - 1. \tag{4.2.6}$$

Thus, for every X, $\Lambda_n(X) \to \infty$ as $n \to \infty$.[†] This fact has the somewhat disappointing consequence that, given X, there exists a continuous function, f, such that $L_n(f, X)$ does *not* converge uniformly to f on I as $n \to \infty$. This follows from

THEOREM 4.3.[1]   *Given X, there exists $f^* \in C(I)$ such that*

$$\overline{\lim_{n \to \infty}} \| L_n(f^*, X) \| = \infty. \tag{4.2.7}$$

Proof.   Suppose that the theorem is false; that is, suppose that

$$\| L_n(f, X) \| < M, \tag{4.2.8}$$

all $f \in C(I)$, $n = 0, 1, 2, \ldots$. Suppose, also, that

$$\Lambda_n(X) = \lambda_n(x_0) = \sum_{i=1}^{n+1} |l_i(x_0)|.$$

Let f_n be a continuous function that satisfies $\| f_n \| = 1$, and $f_n(x_i^{(n+1)}) = \text{sgn } l_i(x_0)$. Then

$$\Lambda_n(X) \geq \| L_n(f_n, X) \| \geq \left| \sum_{i=1}^{n+1} f_n(x_i^{(n+1)}) l_i(x_0) \right| = \Lambda_n(X),$$

and so

$$\| L_n(f_n, X) \| = \Lambda_n(X).$$

Let $t_i > 0$, $i = 1, 2, \ldots$ be chosen so that

$$\sum_{i=1}^{\infty} t_i < 1.$$

Consider

$$f(x) = \sum_{i=1}^{\infty} t_i f_{n_i}(x), \tag{4.2.9}$$

where the sequence of integers $n_1, n_2, \ldots$ will be chosen in a moment. By the Weierstrass M-test, the series in (4.2.9) is uniformly convergent, and hence $f \in C(I)$.

[†] Indeed, using more delicate methods, Erdös [1] has shown that there is a positive constant, c, such that $\Lambda_n(X) > (2/\pi) \log n - c$.

We now choose $n_1, n_2, \ldots$ as follows. Choose $n_1 = 1$, and suppose that n_i is chosen for $i < p$. Put

$$S_{p-1} = \sum_{i=1}^{p-1} t_i f_{n_i}.$$

Let n_p be chosen so large that

$$M < t_p \frac{\Lambda_{n_p}}{2}, \tag{4.2.10}$$

which can be done, in view of (4.2.6). We then have

$$L_{n_p}(f) = L_{n_p}(S_{p-1}) + L_{n_p}(t_p f_{n_p}) + L_{n_p}\left(\sum_{i=p+1}^{\infty} t_i f_{n_i} \right).$$

Because of (4.2.8) and (4.2.10), we conclude that

$$\left\| L_{n_p}\left(S_{p-1} + \sum_{i=p+1}^{\infty} t_i f_{n_i} \right) \right\| < M < t_p \frac{\Lambda_{n_p}}{2}.$$

Let $y \in I$ be chosen so that

$$|L_{n_p}(f_{n_p}; y)| = \|L_{n_p}(f_{n_p})\| = \Lambda_{n_p}.$$

Then

$$\|L_{n_p}(f)\| \geq |L_{n_p}(f; y)| \geq |L_{n_p}(t_p f_{n_p}; y)| - \frac{t_p \Lambda_{n_p}}{2} = \frac{t_p \Lambda_{n_p}}{2}. \tag{4.2.11}$$

Now, we further require of n_p that

$$t_p \Lambda_{n_p} \to \infty \qquad \text{as} \quad p \to \infty;$$

then (4.2.11) contradicts (4.2.8), and the theorem is true. ∎

Since f^* is continuous on I, and hence bounded, it follows from (4.2.7) that $L_n(f^*; X)$ does not converge uniformly to f^* on I as $n \to \infty$. Another way of expressing this fact is that there is no "universally effective" set of nodes for polynomial interpolation to the continuous functions. As counterweight to this negative conclusion we have

THEOREM 4.4.   *Given $f \in C(I)$, there exists X such that $L_n(f, X)$ converges uniformly to f on I.*

Proof.   Let $p_n^* \in P_n$ be the best uniform approximation to f on I. By Theorem 1.7 there exist $n + 2$ points of I, with the property that $f - p^*$ has opposite signs at any two consecutive points. Thus, $f - p^*$ has $n + 1$ zeros in I; call them $x_1^{(n+1)}, \ldots, x_{n+1}^{(n+1)}$. Let X be the array of nodes consisting of these zeros. Then $L_n(f, X) = p_n^*$, and the theorem follows from the Weierstrass Approximation Theorem (or Theorem 1.4). ∎

Theorem 4.2 shows that Λ_n has *at least* logarithmic growth. We study next a set of nodes for which Λ_n has *at most* logarithmic growth. These nodes are the zeros of the Chebyshev polynomials. Let T denote the triangular array of nodes whose nth row is defined by

$$x_i^{(n)} = -\cos \frac{(2i-1)\pi}{2n}, \qquad i = 1, \ldots, n.$$

THEOREM 4.5. *If $n \geq 1$,*

$$\Lambda_n(T) < \frac{2}{\pi} \log n + 4. \tag{4.2.12}$$

Proof. The proof is rather long and falls into two parts. First, we propose to show that $\Lambda_n(T) = \lambda_n(T; 1)$ (following Powell [2]). Then we shall show that $\lambda_n(T; 1)$ satisfies (4.2.12).

We see in Exercise 4.4 that

$$\lambda_n(T; x) = \frac{|\cos(n+1)\theta|}{2(n+1)} \sum_{j=1}^{n+1} \left| \cot \frac{\theta + \theta_j}{2} - \cot \frac{\theta - \theta_j}{2} \right|,$$

where $x = \cos \theta$ and

$$\theta_j = \frac{(2j-1)\pi}{2(n+1)}, \qquad j = 1, \ldots, n+1.$$

We wish to show that

$$\chi(\theta) = |\cos(n+1)\theta| \sum_{j=1}^{n+1} \left| \cot \frac{\theta + \theta_j}{2} - \cot \frac{\theta - \theta_j}{2} \right| \tag{4.2.13}$$

attains its maximum value on $[0, \pi]$ at $\theta = 0$. If we subtract $k\pi/(n+1)$ from θ, where k is any integer satisfying $1 \leq k \leq n+1$, only the summands in (4.2.13) change. Moreover, precisely the same set of $2(n+1)$ cotangents recur in the summation in (4.2.13), although the change in θ causes them to be paired differently. Suppose that

$$\max_{0 \leq \theta \leq \pi} \chi(\theta) = \chi(\theta')$$

and for some $\mu = 0, \ldots, n$,

$$\frac{(2\mu - 1)\pi}{2(n+1)} < \theta' \leq \frac{(2\mu + 1)\pi}{2(n+1)}, \tag{4.2.14}$$

then

$$\left| \theta' - \frac{\mu\pi}{n+1} \right| \leq \frac{\pi}{2(n+1)}. \tag{4.2.15}$$

If we put $\theta'' = |\theta' - \mu\pi/(n + 1)|$, then

$$0 \le {}'\theta'' \le \frac{\pi}{2(n + 1)}. \qquad (4.2.16)$$

But if (4.2.16) holds, then, for $j = 1, \ldots, n + 1$,

$$0 \le \frac{\theta'' + \theta_j}{2} \le \frac{\pi}{2} \quad \text{and} \quad 0 \le -\frac{\theta'' - \theta_j}{2} \le \frac{\pi}{2}.$$

Hence, both $\cot (\theta'' + \theta_j)/2$ and $-\cot (\theta'' - \theta_j)/2$ are *nonnegative*, and the same is true of

$$\cot \frac{\theta'' + \theta_j}{2} - \cot \frac{\theta'' - \theta_j}{2}.$$

Thus the value of $\chi(\theta'')$ is the same for all possible pairings of cotangents in (4.2.13). Since $\chi(\theta'') = \chi(-\theta'')$ (χ is an even function of θ), we have

$$\chi(\theta') = \chi(\theta''),$$

and we have established that

$$\max_{0 \le \theta \le \pi} \chi(\theta) = \max_{0 \le \theta \le [\pi/2(n+1)]} \chi(\theta).$$

From (4.2.13) and Exercise 4.4, we have

$$\chi(\theta) = 2|\cos (n + 1)\theta| \sum_{j=1}^{n+1} \frac{\sin \theta_j}{|\cos \theta - \cos \theta_j|},$$

and so, if

$$0 \le \theta \le \frac{\pi}{2(n + 1)}, \qquad (4.2.17)$$

then we have

$$\chi(\theta) = 2 \sum_{j=1}^{n+1} \sin \theta_j \frac{\cos (n + 1)\theta}{\cos \theta - \cos \theta_j} = 2^{n+1} \sum_{j=1}^{n+1} \left\{ \sin \theta_j \prod_{i=1, i \ne j}^{n+1} (\cos \theta - \cos \theta_i) \right\}. \qquad (4.2.18)$$

But each bracket in (4.2.18) contains a nonnegative function that is monotone decreasing in the interval (4.2.17); hence the sum of these functions (multiplied by 2^{n+1}), $\chi(\theta)$, attains its maximum at $\theta = 0$. Thus, we have established that

$$\Lambda_n(T) = \lambda_n(T; 1) = \frac{1}{n + 1} \sum_{j=1}^{n+1} \cot \frac{\theta_j}{2}. \qquad (4.2.19)$$

This concludes the first part of the proof. To continue we consider

$$\pi \Lambda_n(T) = \frac{\pi}{n + 1} \sum_{j=1}^{n+1} \left[\cot \frac{\theta_j}{2} - \frac{1}{\theta_j/2} \right] + \frac{\pi}{n + 1} \sum_{j=1}^{n+1} \frac{1}{\theta_j/2}. \qquad (4.2.20)$$

The function

$$h(\theta) = \cot \frac{\theta}{2} - \frac{1}{\theta/2}, \qquad 0 \le \theta \le \pi \qquad (4.2.21)$$

has as its derivative

$$h'(\theta) = \frac{1}{2} \left[\frac{1}{(\theta/2)^2} - \frac{1}{\sin^2 (\theta/2)} \right] \le 0$$

(recall Lemma 1.7), and hence is monotone decreasing in $0 \le \theta \le \pi$. But $h(0) = 0$, hence

$$\frac{\pi}{n+1} \sum_{j=1}^{n+1} \left[\cot \frac{\theta_j}{2} - \frac{1}{\theta_j/2} \right] \le 0. \qquad (4.2.22)$$

Furthermore,

$$\frac{\pi}{n+1} \sum_{j=1}^{n+1} \frac{1}{\theta_j/2} = 4 \left(\sum_{j=1}^{2(n+1)} \frac{1}{j} - \frac{1}{2} \sum_{j=1}^{n+1} \frac{1}{j} \right). \qquad (4.2.23)$$

By an argument similar to that given on p. 60 it is easy to show that

$$\sum_{j=1}^{k} \frac{1}{j} > \log (k+1) \qquad \text{and} \qquad \sum_{j=1}^{k} \frac{1}{j} < 1 + \log k.$$

Upon using these inequalities in (4.2.23) we obtain, for $n \ge 1$,

$$\frac{\pi}{n+1} \sum_{j=1}^{n+1} \frac{1}{\theta_j/2} < 4 \left(1 + \log 2 + \log (n+1) - \frac{1}{2} \log (n+2) \right)$$

$$< 4 \left(1 + \log 4 + \frac{1}{2} \log n \right) = 4(1 + \log 4) + 2 \log n.$$

$$(4.2.24)$$

Combining (4.2.24) and (4.2.22) yields

$$\Lambda_n(T) < \frac{2}{\pi} \log n + \frac{4}{\pi} (1 + \log 4) < \frac{2}{\pi} \log n + 4. \quad \blacksquare$$

When Theorem 4.4 is compared to Theorem 4.2 and particularly to the sharper version of (4.2.6) mentioned in the footnote on p. 91, it becomes clear in view of Theorem 4.1 that the Chebyshev nodes, T, are a particularly good choice for interpolation if good uniform approximation is desired (see also Exercise 4.3). However, T is not an optimal set of nodes; that is, there exists some set of nodes X such that

$$\Lambda_n(X) < \Lambda_n(T), \qquad n \ge 1.$$

This observation follows from $\Lambda_n(T) = \lambda_n(T; 1)$ (cf. Luttmann and Rivlin [1]).

It is often convenient to interpolate in equally spaced points, and we turn next to a consideration of the resulting Lebesgue constants. Let E denote the array of nodes defined by

$$x_j^{(n)} = -1 + \frac{2(j-1)}{n-1}, \qquad j = 1, \ldots, n. \qquad (4.2.25)$$

Let us suppose that n is odd, $n = 2m + 1$; then the set of nodes (4.2.25) is $\pm j/m, j = 0, \ldots, m$ and

$$\lambda_{n-1}(E; x) = \prod_{|j| \leq m} \left| x - \frac{j}{m} \right| \cdot \sum_{i=-m}^{m} \frac{m^{2m}}{(2m)!} \binom{2m}{m-i} \frac{1}{|x - i/m|}. \qquad (4.2.26)$$

Since λ_{n-1} is an even function of x, we restrict our attention to $0 < x < 1$. (0 is a node, since n is odd, and 1 is always a node.)

An upper bound is easily obtained. Since $|x - j/m| \leq 1$, $j \geq 0$, $|x - j/m| \leq 2, j < 0$, we have

$$\lambda_{n-1}(E; x) \leq \frac{2^m m^{2m}}{(2m)!} \sum_{i=-m}^{m} \binom{2m}{m-i} = \frac{2^{3m} m^{2m}}{(2m)!}. \qquad (4.2.27)$$

To obtain a lower bound, we observe that, since the terms of the sum in (4.2.26) are positive,

$$\lambda_{n-1}(E; x) > \frac{m^{2m}}{m!m!} \prod_{1 \leq |j| \leq m} \left| x - \frac{j}{m} \right| = \frac{1}{m!m!} \prod_{1 \leq |j| \leq m} |mx - j|.$$

(Take the term $i = 0$ alone).

Consider

$$A = \int_{1-(1/m)}^{1} \prod_{1 \leq |j| \leq m} (mx - j) \, dx,$$

and note that $A < 0$. If we make the change of variables $t = m - mx$, we obtain

$$A = \frac{1}{m} \int_0^1 \prod_{1 \leq |j| \leq m} (m - j - t) \, dt$$

$$= -\frac{1}{m} \int_0^1 [(2m - t)(2m - 1 - t) \cdots (2 - t)] \left\{ \frac{(1-t)t}{m-t} \right\} dt$$

$$< -\frac{(2m-1)!}{m} \int_0^1 \frac{(1-t)t}{m-t} \, dt < -\frac{(2m-1)!}{m^2} \int_0^1 (1-t)t \, dt$$

$$< -\frac{(2m-1)!}{6m^2}.$$

Therefore,

$$\max_{1-(1/m) \leq x \leq 1} \lambda_{n-1}(E; x) \geq \frac{1}{m} \int_{1-(1/m)}^{1} \lambda_{n-1}(E; x) \, dx \geq \frac{1}{m \cdot m!m!} |A| > \frac{(2m-1)!}{6m^3 m!m!},$$

and hence

$$\Lambda_{n-1}(E) > \frac{(2m - 1)!}{6m^3 m! m!}. \qquad (4.2.29)$$

We can extract useful bounds from (4.2.27) and (4.2.29) if we have an estimate for $k!$ for large k. The reader familiar with Stirling's formula will see its use here. We shall next derive a weak form of Stirling's formula as follows:

LEMMA 4.2. *If $k > 2$, then we have*

$$k^k e^{-k} \cdot e < k! < k^k e^{-k} \sqrt{k} \cdot e. \qquad (4.2.30)$$

Proof. On the one hand, $y = \log x$ is monotone increasing for $x > 0$;

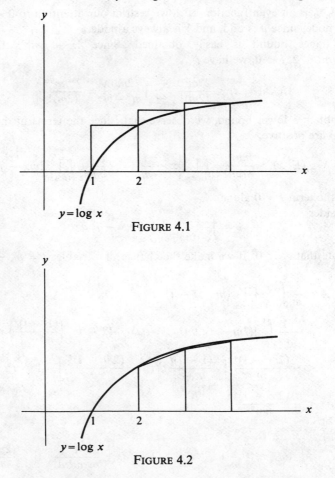

FIGURE 4.1

FIGURE 4.2

hence a comparison of areas in Figure 4.1 indicates that

$$\log 2 + \log 3 + \cdots + \log k > \int_1^k \log x \, dx = k \log k - k + 1.$$

Or,

$$\log k! > (\log k^k) - k + 1,$$

and so

$$k! > k^k e^{-k} \cdot e. \tag{4.2.31}$$

On the other hand, since $y'' = -1/x^2$, $\log x$ is concave for $x > 0$, and as a comparison of areas in Figure 4.2 reveals,

$$\sum_{j=2}^{k-1} \frac{\log j + \log (j+1)}{2} = \frac{\log 2}{2} + \log 3 + \cdots + \log (k-1) + \frac{\log k}{2}$$

$$< \int_2^k \log x \, dx = k \log k - k - 2 \log 2 + 2.$$

Or

$$\log k! < (k + \tfrac{1}{2}) \log k - k - \tfrac{3}{2} \log 2 + 2,$$

so that

$$k! < \sqrt{k} \, k^k e^{-k} \cdot e. \quad \blacksquare \tag{4.2.32}$$

(4.2.30) taken in conjunction with (4.2.27) and (4.2.29) yields

THEOREM 4.6. *If $m \geq 2$, there are constants K_1 and K_2 such that*

$$K_1 (\sqrt{\tfrac{3}{2}})^{2m} < \Lambda_{2m}(E) < K_2 (\sqrt{2} \, e)^{2m}. \tag{4.2.33}$$

A more careful analysis leads to improvement on $\sqrt{\tfrac{3}{2}}$ and $\sqrt{2} \, e$, but the basic fact of exponential growth is contained in (4.2.33).

As a final cautionary note about polynomial interpolations, we shall quote S. Bernstein's result that the sequence of polynomials interpolating the simple function $|x|$ at the equally spaced nodes, E, diverges at all points of $[-1, 1]$ other than $-1, 0, 1$.

THEOREM 4.7. *If $t \in I$ and $0 < |t| < 1$, then*

$$L_{n_k}(|x|, E; t) \to \infty$$

for some infinite sequence of positive integers, $\{n_k\}$.

For the proof see Natanson [1, Vol. III, p. 30]. Theorem 4.7 is an interesting complement to Theorem 4.3, which, for say, $X = E$, admits the possibility that $L_n(f, E)$ may converge at each point of I, possibly to f. Theorem 4.7 exhibits a particularly simple function for which the interpolating polynomials on E diverge throughout I, except at a few points.

In view of (4.1.10) a triangular array of nodes which had minimal Lebesgue constant of order k, for each k, would be of interest. It is not hard to see that there exists such an *extremal* array. For in seeking an extremal array it suffices to restrict our attention to X such that $\Lambda_k(X) \leq ((2/\pi) \log k) + 4 = A_k$, in view of (4.2.12). Let $x_1,...,x_{k+1}$ be the $k + 1^{st}$ row of X. Then $A_k \geq \Lambda_k(X) \geq \| l_j \|$, $j = 1,...,k + 1$ and since

$$\max_{-1 \leq x \leq 1} \prod_{i=1, i \neq j}^{k+1} | x - x_i | \geq \frac{1}{(k + 2)^k} \, , j = 1,...,k + 1$$

we see that

$$\prod_{i=1, i \neq j}^{k+1} | x_j - x_i | \geq \frac{1}{A_k (k + 2)^k} \, , j = 1,...,k + 1.$$

In particular, then, $(x_{j+1} - x_j) \geq (2^k A_k (k + 2)^k)^{-1} = B_k > 0$, $j = 1,...,k$. $\Lambda_k(X)$ is a *continuous* function of $(x_1,...,x_{k+1})$ on the compact set in $k+1$ space defined by $-1 \leq x_1 < x_2 < ... < x_{k+1} \leq 1$ and $(x_{j+1} - x_j) \geq B_k$, $j = 1,...,k$, hence attains its minimum there. Let $(x_1^*,...,x_{k+1}^*)$ be a point at which this minimum is attained. Clearly $-1 \leq x_1^* < x_2^* < ... < x_{k+1}^* \leq 1$. By choosing such a point for each $k = 1,2,...$ we obtain an extremal array, X^*, i.e., $\Lambda_k(X) \geq \Lambda_k(X^*)$, $k = 1,2,...$, for every X.

We observe next that if X is any array and its $k + 1^{st}$ row is $x_1,...,x_{k+1}$ there is a unique a and b such that $ax_1 + b = -1$ and $ax_{k+1} + b = 1$. In this way we find that there is an array X' whose $k + 1^{st}$ row consists of $x_i' = ax_i + b$, i=1,...,k+1, where $x_1' = -1$ $x_{k+1}' = 1$, and, if we put $x' = ax + b$

$$\Lambda_k(X') = \max_{-1 \leq x' \leq 1} \sum_{j=1}^{k+1} \prod_{i=1, i \neq j}^{k+1} | \frac{x' - x_i'}{x_j' - x_i'} |$$

$$= \max_{x_1 \leq x \leq x_{k+1}} \sum_{j=1}^{k+1} \prod_{i=1, i \neq j}^{k+1} | \frac{x - x_i}{x_j - x_i} |$$

$$\leq \Lambda_k(X). \tag{4.2.34}$$

If we put $X = X^*$ in (4.2.34) we see that there is always an extremal array whose every row (beyond the first) includes ± 1 as nodes. Indeed, in view of (4.2.34) we assume, in the rest of this discussion, that every row (beyond the first) of every X includes ± 1 as nodes.

Suppose $k \geq 2$. It can be shown (Luttmann and Rivlin [1]) that $\lambda_k(X;x)$ (see (4.1.9)) has a single maximum in each interval (x_j, x_{j+1}), $j = 1, ..., k + 1$. Let

$$M_{j,k} = M_{j,k}(X) = \max_{x_j \leq x \leq x_{j+1}} \lambda_k(X;x), \quad j = 1, ..., k.$$

S. Bernstein (Izv. Akad. Nauk SSSR 1 (1931), 1025-1050) conjectured that if

$$M_{1,k} = M_{2,k} = ... = M_{k,k}, \quad k = 2,3,... \tag{4.2.35}$$

then X is an extremal array. Erdös (Acta Math. Acad. Sci. Hungar. 9 (1958), 381-288) amplified Bernstein's conjecture as follows: There is a unique array for which (4.2.35) holds, and for any array X

$$\min_{1 \leq j \leq k} M_{j,k}(X) \leq \Lambda_k(X^*). \tag{4.2.36}$$

These conjectures were proved by Kilgore, and de Boor and Pincus (J. Approx. Theory **24**, (1978), 273-288, 289-303). However the nodes of the extremal array are not given explicitly in that work. Indeed, Brutman (SIAM J. Numer. Anal. **15** (1978), 694-704) has shown that if the array T (p. 93) is "expanded" so as to include ± 1 as nodes, i.e., a new array, T', is formed by multiplying each entry in the nth row of T by $\sec(2\pi/n)$ then

$$\Lambda_k(T') \leq \min_{1 \leq j \leq k} M_{j,k}(T') + .201.$$

Thus in view of (4.2.36) the readily available expanded Chebyshev nodes are, for all practical purposes, as useful as the optimal nodes.

4.3 Interpolating Polynomials as Least-Squares and Least-First-Power Approximations

Thus far, we have considered the effectiveness of interpolating polynomials as uniform approximations. Now we turn to a brief study of their effectiveness in the sense of least-squares and least-first-powers.

Let $w(x)$ be a nonnegative weight function on I, and let $\{p_j\}_{j=0}^{\infty}$ be a sequence of orthogonal polynomials with respect to $w(x)$; that is, $p_k \in P_k$ and

$$\int_{-1}^{1} p_i(x)p_j(x)w(x)\,dx = 0, \qquad i \neq j$$

(see Chapter 2). According to Exercise 2.6, p_k has k distinct zeros in $(-1, 1)$, $k = 0, 1, 2, \ldots$. Let W denote the triangular array of nodes whose kth row consists of the zeros of p_k, so that W is uniquely determined by $w(x)$. We are going to study interpolation in the nodes of W, and will be suppressing the W in the notation.

LEMMA 4.3

$$\int_{-1}^{1} l_i^{(k)}(x)l_j^{(k)}(x)w(x)\,dx = 0; \qquad i \neq j, \qquad i, j = 1, \ldots, k+1. \quad (4.3.1)$$

Proof. $l_i = c_i p_{k+1}/(x - x_i)$, $i = 1, \ldots, k+1$; hence

$$l_i l_j = c_i p_{k+1} \frac{l_j}{(x - x_i)}. \quad (4.3.2)$$

But if $i \neq j$, then $l_j/(x - x_i) \in P_{k-1}$, to which, therefore, p_{k+1} is orthogonal with respect to $w(x)$. Hence, (4.3.2) implies (4.3.1). ∎

LEMMA 4.4

$$\int_{-1}^{1} l_i^{(k)}(x)w(x)\,dx = \int_{-1}^{1} [l_i^{(k)}(x)]^2 w(x)\,dx, \qquad i = 1, \ldots, k+1.$$

Proof

$$\int_{-1}^{1} l_i(x)(l_i(x) - 1)w(x)\,dx = c_i \int_{-1}^{1} p_{k+1}(x)\frac{l_i(x) - 1}{x - x_i}\,w(x)\,dx = 0,$$

for, since $l_i(x_i) = 1$, $(l_i - 1)/(x - x_i) \in P_{k-1}$ and p_{k+1} is orthogonal to all polynomials in P_{k-1}. ∎

THEOREM 4.8.   If $f \in C(I)$, then

$$\int_{-1}^{1} [f(x) - L_n(f, W; x)]^2 w(x)\,dx \leq 4E_n^2(f; I)\int_{-1}^{1} w(x)\,dx.$$

Proof. Let $p \in P_n$ be the best uniform approximation to f on I, so that $\|f - p\| = E_n(f; I)$; then

$$\int_{-1}^{1} [f(x) - p(x)]^2 w(x)\,dx \leq E_n^2 \int_{-1}^{1} w(x)\,dx. \quad (4.3.3)$$

But since $L_n(p) = p$, we have

$$L_n(f, W; x) - p(x) = L_n(f - p, W; x) = \sum_{i=1}^{n+1} [f(x_i^{(n+1)}) - p(x_i^{(n+1)})]l_i^{(n)}(x),$$

and hence, from Lemma 4.3,

$$\int_{-1}^{1} [L_n(x) - p(x)]^2 w(x)\, dx$$

$$= \sum_{i=1}^{n+1} \left\{ [f(x_i^{(n+1)}) - p(x_i^{(n+1)})]^2 \int_{-1}^{1} [l_i^{(n)}(x)]^2 w(x)\, dx \right\}$$

$$\leq E_n^2 \sum_{i=1}^{n+1} \int_{-1}^{1} [l_i^{(n)}(x)]^2 w(x)\, dx.$$

After applying Lemma 4.4 and Exercise 4.1, we obtain

$$\int_{-1}^{1} [L_n(x) - p(x)]^2 w(x)\, dx$$

$$\leq E_n^2 \sum_{i=1}^{n+1} \int_{-1}^{1} [l_i^{(n)}(x)]^2 w(x)\, dx = E_n^2 \sum_{i=1}^{n+1} \int_{-1}^{1} l_i^{(n)}(x) w(x)\, dx \tag{4.3.4}$$

$$\leq E_n^2 \int_{-1}^{1} \left[\sum_{i=1}^{n+1} l_i^{(n)}(x) \right] w(x)\, dx = E_n^2 \int_{-1}^{1} w(x)\, dx.$$

Now, for any two real numbers A, B, $(A + B)^2 + (A - B)^2 = 2(A^2 + B^2)$, and so

$$(A + B)^2 \leq 2(A^2 + B^2).$$

Thus

$$(f - L_n)^2 = [(f - p) + (p - L_n)]^2 \leq 2(f - p)^2 + 2(L_n - p)^2,$$

and, from (4.3.3) and (4.3.4), we conclude that

$$\int_{-1}^{1} [f(x) - L_n(x)]^2 w(x)\, dx \leq 4E_n^2 \int_{-1}^{1} w(x)\, dx. \quad \blacksquare$$

Since

$$\int_{-1}^{1} w(x)\, dx \tag{4.3.5}$$

exists, by definition of a weight function, and $E_n \to 0$ as $n \to \infty$ for f continuous, we see that the sequence of polynomials interpolating in the rows of W converges, in the least-squares sense, to f, and the rapidity of convergence can be estimated by using the bounds of E_n (which depend on the smoothness of f) given on p. 23.

COROLLARY 4.8.1. *If*

$$w(x) \geq m > 0, \qquad x \in I,$$

then

$$\int_{-1}^{1} [f(x) - L_n(f, W; x)]^2\, dx \leq \frac{4}{m} E_n^2 \int_{-1}^{1} w(x)\, dx.$$

Theorem 4.8 also leads to some results concerning least-first-power approximation.

COROLLARY 4.8.2

$$\int_{-1}^{1} |f(x) - L_n(f, W; x)| w(x)\, dx \le 2E_n(f; I) \int_{-1}^{1} w(x)\, dx. \qquad (4.3.6)$$

Proof. Apply the Cauchy-Schwarz inequality (Exercise I.2) to

$$\int_{-1}^{1} |f(x) - L_n(x)| w(x)\, dx = \int_{-1}^{1} (|f(x) - L_n(x)| \sqrt{w(x)}) \sqrt{w(x)}\, dx. \quad \blacksquare$$

COROLLARY 4.8.3. *If*

$$\int_{-1}^{1} \frac{1}{w(x)}\, dx$$

exists, then

$$\int_{-1}^{1} |f(x) - L_n(x)|\, dx \le 2E_n \left[\int_{-1}^{1} w(x)\, dx \cdot \int_{-1}^{1} \frac{1}{w(x)}\, dx \right]^{1/2}.$$

Proof. Apply the Cauchy-Schwarz inequality (Exercise I.2) to

$$\int_{-1}^{1} |f(x) - L_n(x)|\, dx = \int_{-1}^{1} (|f(x) - L_n(x)| \sqrt{w(x)}) \left(\frac{1}{\sqrt{w(x)}} \right) dx. \quad \blacksquare$$

Apart from their intrinsic interest, Corollaries 4.8.2 and 4.8.3 provide error bounds for numerical integration schemes of the Gaussian type. This matter is explored further in the exercises.

4.4 Interpolation and Approximation by Splines

Our study of approximation by splines begins with the interpolation problem.

4.4.1 Spline Interpolation

Polynomial interpolation has the drawback of producing approximations that may be excessively oscillatory between the nodes, as we have seen. This oscillatory behavior is somewhat smoothed out when the polynomial is integrated, as Exercise 4.9 testifies. Polynomial interpolation shares with all forms of polynomial approximation the quality of producing an approximation that is analytic and hence amenable to extensive mathematical manipulation. If we abandon the requirement that the approximation (necessarily) be a polynomial, a much more general family of approximating functions that suggests itself is the set of *piecewise* polynomials; i.e., functions that are polynomials, possibly different polynomials in different subdomains of the domain on which we are approximating. The simplest continuous piecewise polynomials are the piecewise linear functions. Such functions have "corners" where two linear pieces meet and, generally, have no derivative at a corner. We shall be studying the approximating properties of the "smooth" piecewise *cubic* functions. Such functions are called (cubic) "splines" for a reason to be mentioned later.[2]

Suppose that X_n denotes the set of real numbers $\{x_0, \ldots, x_n\}$, where

$$a = x_0 < x_1 < \cdots < x_n = b$$

and $f(x)$ is defined on $[a, b]$.

Let $S = S(X_n)$ be the set of all functions $s(X_n; x) = s(x) \in C^2[a, b]$ having the property that, in each interval $[x_i, x_{i+1}]$, $i = 0, \ldots, n-1$, $s(x)$ agrees with a polynomial of degree at most 3. Such functions, s, we call (*cubic*) *splines*. The set of polynomials of degree at most 3, P_3, is clearly a subset of S. We call the points $x_0, \ldots, x_n$ *nodes* (other authors prefer "knots" or "joints"). Our first objective is to determine under what circumstances there exists an interpolating spline, i.e., a function

$$s(f, X_n; x) = s(x) \in S, \tag{4.4.1}$$

which satisfies

$$s(x_i) = f(x_i) = f_i, \qquad i = 0, \ldots, n.$$

To this end, we require a convenient form for cubic polynomials. The reader should verify

LEMMA 4.5.   *If $\alpha < \beta$, the unique $p \in P_3$ that satisfies*

$$\begin{array}{ll} p(\alpha) = u_1, & p(\beta) = u_2, \\ p'(\alpha) = u_1', & p'(\beta) = u_2', \end{array} \tag{4.4.2}$$

is

$$p(x) = u_1 \left[\frac{(x - \beta)^2}{(\beta - \alpha)^2} + 2 \frac{(x - \alpha)(x - \beta)^2}{(\beta - \alpha)^3} \right]$$

$$+ u_2 \left[\frac{(x - \alpha)^2}{(\beta - \alpha)^2} - \frac{2\,(x - \beta)(x - \alpha)^2}{(\beta - \alpha)^3} \right]$$

$$+ u_1' \frac{(x - \alpha)(x - \beta)^2}{(\beta - \alpha)^2} + u_2' \frac{(x - \alpha)^2(x - \beta)}{(\beta - \alpha)^2}. \tag{4.4.3}$$

THEOREM 4.9.   *Given numbers s_0' and s_n', there exists a unique spline satisfying*

$$s(f, X_n; x_i) = f_i, \qquad i = 0, \ldots, n, \tag{4.4.4}$$

and

$$s'(f, X_n; x_i) = s_i', \qquad i = 0, n. \tag{4.4.5}$$

Proof. The strategy of the proof is the following: In the first interval, $[x_0, x_1]$, we choose s to agree with the cubic that has values f_0, f_1 at x_0, x_1, respectively, whose derivative at x_0 is s_0' and whose derivative at x_1 has a value chosen so that the second derivative of this cubic at x_1 agrees with the second derivative at x_1 of the cubic chosen for $[x_1, x_2]$, when this cubic for $[x_1, x_2]$ has values f_1, f_2 at x_1, x_2, respectively, its derivative at x_2 is chosen so as to further this process to $[x_2, x_3]$, etc. The point of the proof is to show that a simultaneous choice of the (first) derivatives at $x_1, \ldots, x_{n-1}$ that carries out this process is possible, and we now proceed to do this.

In $[x_j, x_{j+1}]$, $j = 0, \ldots, n - 1$, let $s(x)$ agree with the cubic, p, which satisfies

$$p(x_j) = f_j, \qquad p(x_{j+1}) = f_{j+1},$$
$$p'(x_j) = s_j', \qquad p'(x_{j+1}) = s_{j+1}',$$

where s_j', $j = 1, \ldots, n - 1$, remain to be determined. Then, by substituting in Lemma 4.5, we obtain, for $x_j \le x \le x_{j+1}, j = 0, \ldots, n - 1$,

$$s(x) = f_j \left[\frac{(x - x_{j+1})^2}{(\Delta x_j)^2} + 2 \frac{(x - x_j)(x - x_{j+1})^2}{(\Delta x_j)^3} \right]$$

$$+ f_{j+1} \left[\frac{(x - x_j)^2}{(\Delta x_j)^2} - 2 \frac{(x - x_{j+1})(x - x_j)^2}{(\Delta x_j)^3} \right]$$

$$+ s_j' \left[\frac{(x - x_j)(x - x_{j+1})^2}{(\Delta x_j)^2} \right] + s_{j+1}' \left[\frac{(x - x_j)^2(x - x_{j+1})}{(\Delta x_j)^2} \right], \quad (4.4.6)$$

where

$$\Delta x_j = x_{j+1} - x_j. \tag{4.4.7}$$

Thus,

$$s''(x_j) = -\frac{6}{(\Delta x_j)^2} f_j + \frac{6}{(\Delta x_j)^2} f_{j+1} - \frac{4}{\Delta x_j} s_j' - \frac{2}{\Delta x_j} s_{j+1}' \tag{4.4.8}$$

and

$$s''(x_{j+1}) = \frac{6}{(\Delta x_j)^2} f_j - \frac{6}{(\Delta x_j)^2} f_{j+1} + \frac{2}{\Delta x_j} s_j' + \frac{4}{\Delta x_j} s_{j+1}'. \tag{4.4.9}$$

Hence, if $i = 1, \ldots, n - 1$, upon equating $s''(x_i)$ as calculated from $[x_{i-1}, x_i]$ and $[x_i, x_{i+1}]$ by means of (4.4.8) and (4.4.9) and writing $\Delta f_j = f_{j+1} - f_j$, we obtain

$$\Delta x_i s_{i-1}' + 2(\Delta x_i + \Delta x_{i-1})s_i' + \Delta x_{i-1} s_{i+1}' = 3 \left[\frac{\Delta x_i}{\Delta x_{i-1}} \Delta f_{i-1} + \frac{\Delta x_{i-1}}{\Delta x_i} \Delta f_i \right],$$

$$\tag{4.4.10}$$

a system of $n - 1$ linear equations for the $n - 1$ unknowns $s_1', s_2', \ldots, s_{n-1}'$. The matrix of the system (4.4.10) has a quite special form. All its entries are nonnegative and it is tridiagonal and diagonally dominant; that is, in each row the diagonal element is greater than (in this case twice as large as) the sum of all the other entries in that row. This enables us to show that the matrix is nonsingular, and hence that (4.4.10) has a unique solution. For suppose that the matrix were singular; then its columns would be linearly dependent vectors in $(n-1)$-space. That is, if $a_1, \ldots, a_{n-1}$ are the column vectors of the matrix, there would exist numbers $\lambda_1, \ldots, \lambda_{n-1}$, not all zero such that

$$\sum_{i=1}^{n-1} \lambda_i a_i = 0. \qquad (4.4.11)$$

Suppose that $|\lambda_j| \geq |\lambda_i|$, $i = 1, \ldots, n-1$, and consider the jth component of the vector

$$\sum_{i=1}^{n-1} \lambda_i a_i.$$

On the one hand, its absolute value satisfies

$$|\lambda_{j-1} \Delta x_j + \lambda_j 2(\Delta x_j + \Delta x_{j-1}) + \lambda_{j+1} \Delta x_{j-1}|$$
$$\geq |\lambda_j| 2(\Delta x_j + \Delta x_{j-1}) - |\lambda_{j-1}| \Delta x_j - |\lambda_{j+1}| \Delta x_{j-1}$$
$$\geq |\lambda_j|(\Delta x_j + \Delta x_{j-1}) > 0,$$

since $\lambda_j \neq 0$, while (4.4.11) implies that it is zero. This contradiction establishes the unique solvability of (4.4.10) and thus proves the theorem. ∎

To find the value of s at some point x in the interval $[x_i, x_{i+1}]$, the values of s_1' and s_{i+1}' as determined by solving the system (4.4.10) are used in (4.4.3). As an illustration of Theorem 4.9 we observe that there exists a unique spline that satisfies

$$c_i(x_j) = \begin{cases} 0, & i \neq j, \\ 1, & i = j, \end{cases} \qquad (4.4.12)$$

$$c_i'(x_0) = c_i'(x_n) = 0, \qquad (4.4.13)$$

for $i = 0, \ldots, n$. We call this set of splines the *fundamental* splines for spline interpolation, in analogy with the fundamental polynomials that played such an important role in our study of polynomial interpolation.

The set of splines, $S(X_n)$, is, clearly, a linear space. Moreover, it has dimension $n + 3$, and the functions x, x^2, $c_0(x), \ldots, c_n(x)$ form a basis for it. To see this, we observe that

$$t(x) = ax + bx^2 + \sum_{i=0}^{n} \alpha_i c_i(x) \in S. \qquad (4.4.14)$$

Also, if $s \in S(X_n)$ and we write $s_i = s(x_i)$, $i = 0, \ldots, n$, $s'_0 = s'(x_0)$, $s'_n = s'(x_n)$ and put

$$q(x) = \frac{1}{2} \frac{s'_0 - s'_n}{x_0 - x_n} x^2 + \frac{s'_0 x_n - s'_n x_0}{x_n - x_0} x,$$

then $q \in P_2$ and $q'(x_0) = s'_0$, $q'(x_n) = s'_n$. If we write $q_i = q(x_i)$, $i = 0, \ldots, n$, the uniqueness proved in Theorem 4.9 implies that

$$s(x) - q(x) = \sum_{i=0}^{n} (s_i - q_i) c_i(x) \quad \text{or} \quad s(x) = q(x) + \sum_{i=0}^{n} (s_i - q_i) c_i(x).$$

Finally, if $t(x)$, as given by (4.4.14), is identically zero, then $t'(x_0) = t'(x_n) = 0$. Hence, from (4.4.13), $a = b = 0$, while $t(x_i) = 0$ implies $\alpha_i = 0$, $i = 0, \ldots, n$. Thus, $x, x^2, c_0, \ldots, c_n$ are linearly independent and form a basis for $S(X_n)$.

Having shown the possibility of spline interpolation, we shall devote the rest of this chapter to a study of the approximating ability of cubic splines.

4.4.2   Some Extremal Properties of Splines

THEOREM 4.10.   *Suppose that $a = x_0 < x_1 < \cdots < x_n = b$ and $f \in C^2[a, b]$. If we take $f_i = f(x_i)$, $i = 0, \ldots, n$ and consider the spline that satisfies*

$$s(x_i) = f_i, \qquad i = 0, \ldots, n, \tag{4.4.15}$$

and

$$s'(a) = f'(a), \qquad s'(b) = f'(b), \tag{4.4.16}$$

then we have

$$\int_a^b [f''(x)]^2 \, dx - \int_a^b [s''(x)]^2 \, dx = \int_a^b [f''(x) - s''(x)]^2 \, dx. \tag{4.4.17}$$

Proof

$$\int_a^b [f''(x) - s''(x)]^2 \, dx = \int_a^b [f''(x)]^2 \, dx - \int_a^b [s''(x)]^2 \, dx$$
$$- 2 \int_a^b s''(x)[f''(x) - s''(x)] \, dx.$$

Integration by parts yields

$$\int_a^b s''(x)[f''(x) - s''(x)] \, dx = s''(x)[f'(x) - s'(x)] \Big|_a^b - \int_a^b s'''(x)[f'(x) - s'(x)] \, dx.$$

The first term on the right is zero by (4.4.16), and since s''' is a constant in each (x_i, x_{i+1}), say $s'''(x) = \alpha_i$, the second term is

$$\sum_{i=0}^{n-1} \alpha_i \int_{x_i}^{x_{i+1}} [f'(x) - s'(x)] \, dx = 0,$$

from (4.4.15).   ∎

Suppose that $u \in C^2[a, b]$ satisfies

$$u(x_i) = f_i, \qquad i = 0, \ldots, n,$$

and

$$u'(a) = f'(a), \qquad u'(b) = f'(b);$$

then we have

COROLLARY 4.10.1.   *If s is the cubic spline defined by (4.4.15) and (4.4.16),*

$$\int_a^b [u''(x)]^2 \, dx \geq \int_a^b [s''(x)]^2 \, dx, \tag{4.4.18}$$

with equality holding only for u = s.

The minimizing property shown in (4.4.18) helps explain the origin of the name "spline" for interpolating piecewise cubics. Engineers have for a long time used thin rods, called splines, to fair curves through given points. The "strain energy" minimized by such splines is proportional, approximately, to the integral of the square of the second derivative of the spline.

THEOREM 4.11.   *With the same hypotheses as Theorem 4.10, we have*

$$\int_a^b [f''(x) - s''(x)]^2 \, dx \leq \int_a^b [f''(x) - v''(x)]^2 \, dx \tag{4.4.19}$$

for all $v \in S(X_n)$. The equality holds in (4.4.19) if and only if $v = s + ax + b$.

Proof.   Given $v \in S(X_n)$, let the function $f - v$ play the role of f in Theorem 4.10. The role of s in that theorem is then assumed by $s - v$, and (4.4.17) tells us that

$$\int_a^b [f''(x) - v''(x)]^2 \, dx - \int_a^b [s''(x) - v''(x)]^2 \, dx = \int_a^b [f''(x) - s''(x)]^2 \, dx. \tag{4.4.20}$$

(4.4.19) now follows, and the final statement is true, since

$$\int_a^b \{[s(x) - v(x)]''\}^2 \, dx = 0 \Leftrightarrow [s(x) - v(x)]'' \equiv 0 \Leftrightarrow s(x) - v(x) = ax + b. \quad \blacksquare$$

Thus, if approximative power is measured in terms of the pseudonorm

$$\int_a^b [g''(x)]^2 \, dx,$$

the interpolating spline is a best approximation.

We turn next to a determination of the behavior of splines as uniform approximations to given functions.

4.4.3   Uniform Approximation by Splines

Suppose, now, that $a = 0$, $b = 1$; i.e., the set $X_n : \{x_0, \ldots, x_n\}$ satisfies $0 = x_0 < x_1 < \cdots < x_n = 1$. Given $f(x)$ defined on $I: [0, 1]$, we wish to consider the problem of how to choose $s \in S(X_n)$ so that $\|f - s\|$ is minimum, where $\|\cdot\|$ is the uniform norm on I. This is the approximation problem with fixed nodes. An even more ambitious and interesting problem would be to vary the set of nodes X_n in I so as to minimize min $\|f - s(X_n)\|$. We will have little to say about this latter problem, but we refer the reader to Rice [1, Vol. 2].

Let us put $\Delta x_i = x_{i+1} - x_i$, $i = 0, \ldots, n - 1$, and define the norm of X_n by

$$\delta = \delta(X_n) = \max_{i = 0, \ldots, n-1} \Delta x_i.$$

We wish to consider best uniform approximation to $f(x)$ on I out of $S(X_n)$. Among the approximating functions is the interpolating spline whose existence was established in Theorem 4.9. We have

THEOREM 4.12 (SHARMA AND MEIR [1]).   *Suppose that $f \in C^2(I)$ and $s \in S(X_n)$ satisfies*

$$s(x_j) = f(x_j), \qquad j = 0, \ldots, n,$$
$$s'(0) = f'(0), \qquad s'(1) = f'(1). \tag{4.4.21}$$

Then, for all $x \in I$,

$$|f^{(r)}(x) - s^{(r)}(x)| \leq 5\delta^{2-r}\omega(f''; I; \delta), \qquad r = 0, 1, 2.$$

Proof.   Let us write s_j for $s(x_j)$, f_j for $f(x_j)$, s'_j for $s'(x_j)$, f'_j for $f'(x_j)$, etc. Then, upon solving (4.4.8) and (4.4.9) for s'_j and s'_{j+1} and then replacing $j+1$ by j in the latter solution and equating the two resulting forms of s'_j, we obtain the following system of linear equations for s''_j, $j = 0, \ldots, n$:

$$\frac{1}{3}\Delta x_0 s''_0 + \frac{1}{6}\Delta x_0 s''_1 = \frac{\Delta f_0}{\Delta x_0} - f'_0,$$

$$\tfrac{1}{6}\Delta x_{j-1}s''_{j-1} + \tfrac{1}{3}(\Delta x_{j-1} + \Delta x_j)s''_j + \tfrac{1}{6}\Delta x_j s''_{j+1}$$

$$= \frac{\Delta x_j f_{j-1} - (\Delta x_{j-1} + \Delta x_j)f_j + \Delta x_{j-1}f_{j+1}}{\Delta x_{j-1} \Delta x_j}, \qquad j = 1, \ldots, n-1,$$

$$\tag{4.4.22}$$

$$\frac{1}{6}\Delta x_{n-1}s''_{n-1} + \frac{1}{3}\Delta x_{n-1}s''_n = -\frac{\Delta f_{n-1}}{\Delta x_{n-1}} + f'_n.$$

The first and last equations are consequences of (4.4.8), (4.4.9), and (4.4.21). Suppose that $j = 1, \ldots, n - 1$ and that

$$L(x) = f_{j+1} \frac{x - x_{j-1}}{x_{j+1} - x_{j-1}} + f_{j-1} \frac{x_{j+1} - x}{x_{j+1} - x_{j-1}}$$

$$= \frac{1}{\Delta x_{j-1} + \Delta x_j} [f_{j+1}(x - x_{j-1}) + f_{j-1}(x_{j+1} - x)]$$

is the linear polynomial interpolating to $f(x)$ at x_{j-1} and x_{j+1}. Hence, from (4.4.28) in Exercise 4.2, there exists ξ_j satisfying $x_{j-1} \leq \xi_j \leq x_{j+1}$ such that

$$f(x_j) - L(x_j) = \frac{(\Delta x_{j-1} + \Delta x_j)f_j - \Delta x_j f_{j-1} - \Delta x_{j-1} f_{j+1}}{\Delta x_j + \Delta x_{j-1}}$$

$$= -\Delta x_{j-1} \Delta x_j \frac{f''(\xi_j)}{2}. \tag{4.4.23}$$

Thus, if we put

$$A_j = s_j'' - f_j'',$$

then we can rewrite Equations (4.4.22) as

$$\tfrac{1}{3} \Delta x_0 A_0 + \tfrac{1}{6} \Delta x_0 A_1$$
$$= \frac{\Delta f_0}{\Delta x_0} - f_0' - \frac{1}{3} \Delta x_0 f_0'' - \frac{1}{6} \Delta x_0 f_1''$$
$$= \tfrac{1}{3} \Delta x_0 [f''(\xi_0) - f_0''] + \tfrac{1}{6} \Delta x_0 [f''(\xi_0) - f_1''], \tag{4.4.24}$$

$$\tfrac{1}{6} \Delta x_{j-1} A_{j-1} + \tfrac{1}{3} (\Delta x_{j-1} + \Delta x_j) A_j + \tfrac{1}{6} \Delta x_j A_{j+1}$$
$$= (\Delta x_j + \Delta x_{j-1}) \frac{f''(\xi_j)}{2} - \frac{1}{6} \Delta x_{j-1} f_{j-1}'' - \frac{1}{3} (\Delta x_{j-1} + \Delta x_j) f_j'' - \frac{1}{6} \Delta x_j f_{j+1}''$$

$$= \tfrac{1}{6} \Delta x_{j-1} [f''(\xi_j) - f_{j-1}''] + \tfrac{1}{3} (\Delta x_{j-1} + \Delta x_j)[f''(\xi_j) - f_j'']$$
$$+ \tfrac{1}{6} \Delta x_j [f''(\xi_j) - f_{j+1}''], \quad j = 1, \ldots, n - 1,$$

$$\tfrac{1}{6} \Delta x_{n-1} A_{n-1} + \tfrac{1}{3} \Delta x_{n-1} A_n$$
$$= -\frac{\Delta f_{n-1}}{\Delta x_{n-1}} + f_n' - \frac{1}{6} \Delta x_{n-1} f_{n-1}'' - \frac{1}{3} \Delta x_{n-1} f_n''$$
$$= \tfrac{1}{6} \Delta x_{n-1} [f''(\xi_n) - f_{n-1}''] + \tfrac{1}{3} \Delta x_{n-1} [f''(\xi_n) - f_n''],$$

where we have used (4.4.23) and $x_0 < \xi_0 < x_1$, $x_{n-1} < \xi_{n-1} < x_n$.

Now the absolute value of the right-hand side in equation j of (4.4.24) $(j = 1, \ldots, n - 1)$ does not exceed

$$\tfrac{1}{3}(\Delta x_{j-1} + \Delta x_j)\omega(f''; I; \delta) + \tfrac{1}{6}(\Delta x_{j-1} + \Delta x_j)\omega(f''; I; 2\delta)$$
$$\leq \tfrac{2}{3}(\Delta x_{j-1} + \Delta x_j)\omega(f''; I; \delta),$$

in view of the fact that $\omega(2\delta) \leq 2\omega(\delta)$ (see the proof of Lemma 1.3 in Chapter

1, p. 15), and the same bound holds for the first and last equation as well, provided that we put $\Delta x_{-1} = \Delta x_n = 0$, as we do.

Now, suppose that

$$\max_{j=0,\ldots,n} |A_j| = |A_k|,$$

and let us consider the kth equation in (4.4.24). We obtain

$$\tfrac{1}{3}(\Delta x_{k-1} + \Delta x_k)|A_k| \leq \tfrac{2}{3}(\Delta x_{k-1} + \Delta x_k)\omega(f''; I; \delta) + \tfrac{1}{6}\Delta x_{k-1}|A_{k-1}|$$
$$+ \tfrac{1}{6}\Delta x_{k+1}|A_{k+1}| \leq \tfrac{2}{3}(\Delta x_{k-1} + \Delta x_k)\omega(f''; I; \delta)$$
$$+ \tfrac{1}{6}(\Delta x_{k-1} + \Delta x_k)|A_k|,$$

from which we conclude that, for $j = 0, \ldots, n$,

$$|f_j'' - s_j''| \leq |A_k| \leq 4\omega(f''; I; \delta).$$

Now $s''(x)$ is linear in each interval $[x_j, x_{j+1}]$, $j = 0, \ldots, n-1$; hence, in each such interval, either

$$s_j'' \leq s''(x) \leq s_{j+1}'' \quad \text{or} \quad s_{j+1}'' \leq s''(x) \leq s_j'',$$

and, hence, either

$$s_j'' - f_j'' + f_j'' - f''(x) \leq s''(x) - f''(x) \leq s_{j+1}'' - f_{j+1}'' + f_{j+1}'' - f''(x)$$

or

$$s_{j+1}'' - f_{j+1}'' + f_{j+1}'' - f''(x) \leq s''(x) - f''(x) \leq s_j'' - f_j'' + f_j'' - f''(x),$$

from which we conclude that

$$|s''(x) - f''(x)| \leq 4\omega(f''; I; \delta) + \omega(f''; I; \delta) \leq 5\omega(f''; I: \delta)$$

holds for all $x \in I$. This proves the theorem for $r = 2$.

Suppose that $x_j \leq x \leq x_{j+1}$. Since $f_j = s_j$ and $f_{j+1} = s_{j+1}$, there exists η_j satisfying $x_j < \eta_j < x_{j+1}$ such that $f'(\eta_j) - s'(\eta_j) = 0$. Hence

$$|f'(x) - s'(x)| = \left| \int_{\eta_j}^x [f''(t) - s''(t)]\, dt \right| \leq 5\omega(f''; I; \delta)\delta,$$

and so

$$|f(x) - s(x)| = \left| \int_{x_j}^x [f'(t) - s'(t)]\, dt \right| \leq 5\omega(f''; I; \delta)\delta^2.$$

These results are valid for $j = 0, \ldots, n-1$, and thus the theorem is proved in its entirety. ∎

This theorem shows that, if $\delta(X_n) \to 0$ as $n \to \infty$, the interpolating spline and its first and second derivatives (which are also piecewise polynomials) converge uniformly to a given function in $C^2(I)$. The technique of proof used here provides similar results for functions in $C^3(I)$ and $C(I)$ (cf. Sharma and

Meir [1]). Of course, we conclude from Theorem 4.12 that, if $f \in C^2(I)$, then we have

$$\min_{s \in S(X_n)} \|f - s\| \leq 5\delta^2 \omega(f''; I; \delta).$$

It is possible to obtain noninterpolating spline approximations, but the degree of approximation by the interpolating spline is in a sense best possible (for more information about both of these observations see Birkhoff [1]).

4.4.4 Least-Squares Approximation by Splines

We now return to a general interval $[a, b]$. Since $S(X_n)$ is a finite dimensional subspace of $C[a, b]$ (cf. p. 107), we conclude from the results of our introductory chapter that there exists a unique least-squares approximation to a given $f \in C[a, b]$ out of $S(X_n)$. The least-squares approximation is characterized exactly as in the polynomial case. We have

THEOREM 4.13.   *Given $f \in C[a, b]$ and a weight function $w(x)$,*

$$\int_a^b [f(x) - s^*(x)]^2 w(x)\, dx < \int_a^b [f(x) - s(x)] w(x)\, dx$$

for some $s^* \in S(X_n)$ and all $s \in S(X_n)$, $s \neq s^*$, if and only if

$$\int_a^b [f(x) - s^*(x)]s(x)w(x)\, dx = 0 \qquad (4.4.25)$$

for all $s \in S$.

The proof is a repetition of that of Theorem 2.1, and we omit it.

If we put $s = x, x^2, c_0, c_1, \ldots, c_n$ successively in (4.4.25), we obtain

$$\int_a^b x^i s^*(x) w(x)\, dx = \int_a^b x^i f(x) w(x)\, dx, \qquad i = 1, 2,$$

$$\int_a^b c_j(x) s^*(x) w(x)\, dx = \int_a^b c_j(x) f(x) w(x)\, dx, \qquad j = 0, \ldots, n,$$

a system of $n + 3$ linear equations (the normal equations) for the $n + 3$ coefficients of s^* (cf. Equation (4.4.14), p. 107). In this fashion, we obtain the least-squares spline approximation. A method of obtaining s^* that is more congenial computationally may be found in Powell [1].

From (4.4.25), we obtain

$$\int_a^b [f(x) - s^*(x)]^2 w(x)\, dx = \int_a^b [f(x) - s^*(x)]f(x)w(x)\, dx$$

$$= \int_a^b [f(x)]^2 w(x)\, dx - \int_a^b [s^*(x)]^2 w(x)\, dx.$$

$$(4.4.26)$$

Of course, $s^*(x) = s^*(X_n; x)$. Thus, (4.4.26) suggests that, if we vary the set of nodes in $[a, b]$, we should do so in such a manner as to *increase*

$$\int_a^b [s^*(X_n; x)]^2 w(x)\, dx.$$

Further details may be found in Powell [1].

Exercises

4.1 Given any X, show that

$$\sum_{j=1}^{k+1} l_j^{(k)}(x) \equiv 1, \qquad k = 0, 1, \ldots,$$

and hence that

$$\lambda_k(X; x) \geq 1, \qquad k = 0, 1, \ldots, x \in I. \qquad (4.4.27)$$

When does equality hold in (4.4.27)?

4.2 Suppose that f has an nth derivative in I that satisfies $\|f^{(n)}\| = M$. Show that, for each $x \in I$,

$$f(x) - L_{n-1}(x) = \frac{(x - x_1^{(n)}) \cdots (x - x_n^{(n)})}{n!} f^{(n)}(\xi) \qquad (4.4.28)$$

for some $\xi(x) \in I$ and, hence, that

$$G_{n-1} = \|f - L_{n-1}\| \leq \|(x - x_1^{(n)}) \cdots (x - x_n^{(n)})\| \frac{M}{n!}. \qquad (4.4.29)$$

[*Hint:* Let $h(x)/g(x) = [f(x) - L_{n-1}(x)]/[(x - x_1^{(n)}) \cdots (x - x_n^{(n)})]$. If $x \neq x_i^{(n)}$, then $h(x)g(t) - g(x)h(t)$, as a function of t, has $n + 1$ distinct zeros $x, x_1^{(n)}, \ldots, x_n^{(n)}$ in I. Repeated application of Rolle's Theorem implies that $h(x)g^{(n)}(t) - g(x)h^{(n)}(t) = 0$ at $t = \xi \in I$.]

4.3 Show that, if M is as given in Exercise 4.2, the right-hand side of (4.4.29) is least when $x_1^{(n)}, \ldots, x_n^{(n)}$ are the zeros of the Chebyshev polynomial, $T_n(x)$. What is the right-hand side of (4.4.29) in this case?

4.4 Let T be the matrix of nodes whose kth row consists of the zeros of $T_k(x)$, $k = 1, 2, \ldots$; that is

$$x_j^{(k)} = -\cos \frac{(2j - 1)\pi}{2k}, \qquad j = 1, \ldots, k.$$

Show that

$$\lambda_{k-1}(T; x) = \frac{|\cos k\theta|}{k} \sum_{j=1}^{k} \frac{\sin \theta_j}{|\cos \theta - \cos \theta_j|},$$

and, hence, that also,

$$\lambda_{k-1}(T; x) = \frac{|\cos k\theta|}{2k} \sum_{j=1}^{k} \left| \cot \frac{\theta + \theta_j}{2} - \cot \frac{\theta - \theta_j}{2} \right|,$$

where $x = \cos \theta$ and

$$\theta_j = \frac{(2j-1)\pi}{2k}, \quad j = 1, \ldots, k.$$

4.5 If we write c_j^+ for $\cot (\theta + \theta_j)/2$ and c_j^- for $\cot (\theta - \theta_j)/2$, show that, if θ is replaced by $\theta - k\pi/(n+1)$, $k = 1, 2, \ldots, n+1$, then

$$\begin{aligned}
c_j^+ &\to c_{j-k}^+, & j &= k+1, \ldots, n+1, \\
c_j^+ &\to c_{k+1-j}^-, & j &= 1, \ldots, k, \\
c_j^- &\to c_{j+k}^-, & j &= 1, \ldots, n+1-k, \\
c_j^- &\to c_{2n+3-j-k}^+, & j &= n+2-k, \ldots, n+1,
\end{aligned}$$

thereby verifying the assertion made in the proof of Theorem 4.4.

4.6 Show that

$$\lim_{n \to \infty} \left[\Lambda_n(T) - \left(\frac{2}{\pi} \log n + \frac{2}{\pi} \left(\log \frac{8}{\pi} + \gamma \right) \right) \right] = 0,$$

where γ is Euler's constant, $\gamma = 0.5772\ldots$; that is,

$$\gamma = \lim_{k \to \infty} \left[\sum_{j=1}^{k-1} \frac{1}{j} - \log k \right].$$

[*Hint:* Write Λ_n in the form of (4.2.20). Then note that the first sum in (4.2.20) is a Riemann sum for the integral

$$\int_0^\pi h(\theta) \, d\theta = 2 \log \frac{2}{\pi},$$

where $h(\theta)$ is as defined in (4.2.21).]

4.7 Show that, if f defined on I satisfies (the Dini-Lipschitz condition)

$$\lim_{\delta \to 0} \omega(f; \delta) \log \delta = 0, \tag{4.4.30}$$

then $L_n(f, T)$ converges uniformly to f on I. In particular, (4.4.30) surely holds if f has a bounded derivative on I.

4.8   Show that

$$\lim_{n \to \infty} L_{2n-1}(|x|, E; 0) = 0$$

and, hence, that

$$\lim_{n \to \infty} L_n(|x|, E; 0) = 0.$$

[*Hint:* The nodes are $\pm (2j - 1)/(2n - 1), j = 1, \ldots, n.$]
   When

$$I = \int_{-1}^{1} f(x)\, dx$$

is not readily available, one often approximates the integral (this procedure is called *numerical integration*) by choosing some set of nodes, X, and calculating

$$I_n(X) = \int_{-1}^{1} L_n(f, X; x)\, dx.$$

It is evident that

$$I_n(X) = \sum_{i=1}^{n+1} f(x_i^{(n+1)}) \int_{-1}^{1} l_i^{(n)}(x)\, dx,$$

that is, $I_n(X)$ is obtained by sampling f at the points of the $(n + 1)$st row of X and weighting each sample appropriately. A numerical integration scheme in which I is approximated by I_n is called a *Lagrangian* or interpolatory method. In particular, if $X = W$, where W is the matrix of nodes consisting of the zeros of the orthogonal polynomials with respect to a weight function w, the method is of Gaussian type, the name Gaussian, or Gaussian Quadrature being reserved for the case $w \equiv 1$, the kth row of W then consisting of the zeros of P_k, the Legendre polynomial (for more details, see Davis and Rabinowitz [1]).
   4.9   If W is given, show that

$$|I - I_n(W)| \le 2E_n \left[\int_{-1}^{1} w(x)\, dx \cdot \int_{-1}^{1} \frac{1}{w(x)}\, dx \right]^{1/2},$$

so that, for Gaussian Quadrature, $W = W_0$,

$$|I - I_n(W_0)| \le 4E_n.$$

   4.10   If T is the array of Chebyshev nodes, show that

$$|I - I_n(T)| \le \sqrt{2}\, \pi E_n.$$

   Just as interpolating polynomials provide a method for numerical integration, they can be used for numerical differentiation, that is, approximating the derivatives of a given function. Suppose that $f \in C^k[-1, 1]$; the idea is to approximate $D^k f$ by $D^k L_n(f, X)$, where X is an array of nodes as defined in (4.1.4). Thus

$$D^k L_n(f, X) = \sum_{i=1}^{n+1} f(x_i) D^k l_i(x).$$

Let us put

$$\Lambda_n^{(k)}(X) = \max_{-1 \leq x \leq 1} \sum_{i=1}^{n+1} |D^k l_i(x)|,$$

and

$$\mathscr{L}_n^{(k)} = \inf_X \Lambda_n^{(k)}(X).$$

4.11   Show that, if k does not exceed n,

$$\| D^k f - D^k L_n(f, X) \| \leq \Lambda_n^{(k)}(X) E_n(f; [-1, 1]) + \| D^k f - D^k p \|,$$

where p is the best uniform approximation to f on $[-1, 1]$ out of P_n and $\| \cdot \|$ is the uniform norm.

[*Hint:* Mimic the proof of Theorem 4.1.]

4.12   Show that Exercise 4.11 implies that, for $0 \leq k \leq n$,

$$\| D^k f - D^k L_n(f, X) \| \leq E_{n-k}(D^k f) \left[1 + \frac{C}{n^k} \Lambda_n^{(k)}(X) \right],$$

where $C = 6^k e^k (1 + k)^{-1}$.

[*Hint:* Recall the proof of Theorem 1.5.]

4.13   Show that, if $1 \leq k \leq n$,

$$\mathscr{L}_n^{(k)} \geq D^k T_n(1).$$

The value of $D^k T_n(1) = T_n^{(k)}(1)$ is given in (1.2.22).

[*Hint:* For any X,

$$|D^k T_n(x)| \leq \left| \sum_{i=1}^{n+1} T_n(x_i) D^k l_i(x) \right| \leq \Lambda_n^{(k)}(X).]$$

4.14 (Berman)   Show that, if $1 \leq k \leq n$,

$$\mathscr{L}_n^{(k)} = \Lambda_n^{(k)}(U) = D^k T_n(1).$$

This result may be interpreted as saying that the extrema of $T_n(x)$ nodes, U, are an optimal set for numerical differentiation. Compare the discussion on p. 96.

[*Hint:* Suppose that

$$\Lambda_n^{(k)}(U) = \sum_{i=1}^{n+1} |D^k l_i(x^*)| .]$$

Consider the polynomial

$$p(x) = \sum_{i=1}^{n+1} [\text{sgn } D^k l_i(x^*)] l_i(x).$$

Show that the result of Duffin and Schaeffer, mentioned on p. 32, is applicable to $p(x)$ and yields

$$|D^k p(x)| \le D^k T_n(1), \qquad -1 \le x \le 1.]$$

4.15   Show that there is a unique cubic spline satisfying

$$s(x_i) = f_i, \qquad i = 0, \dots, n \qquad \text{and} \qquad s''(x_i) = 0, \qquad i = 0, n.$$

4.16   Suppose that $n > 3$ and $s(x) = 0$ for x outside (x_i, x_{i+3}) for some i, $0 \le i \le n - 3$. Show that $s(x) \equiv 0$.

4.17   Show that, if $n > 3$, there exists $s(x) \in S$, not identically zero, which is zero outside (x_i, x_{i+4}) for each $i = 0, \dots, n - 4$.

[*Hint:* Define

$$x_+^k = \begin{cases} x^k, & x \ge 0, \\ 0, & x < 0, \end{cases}$$

for nonnegative integers k. Consider the function

$$\sum_{j=i}^{i+4} \frac{(x_j - x)_+^3}{\omega'(x_j)},$$

where $\omega(x) = (x - x_i) \cdots (x - x_{i+4}).]$

4.18   If

$$s(x) \in C^3[a, b],$$

show that $s \in P_3$.

4.19   Show that every $s \in S(X_n)$ has the representation

$$s(x) = p(x) + \sum_{i=1}^{n-1} c_i (x - x_i)_+^3$$

for some choice of c_i and $p \in P_3$.

[*Hint:* Let S_i denote the jump in the third derivative of s at x_i, $i = 1, \dots,$ $n - 1$. Examine

$$s(x) - \frac{1}{6} \sum_{i=1}^{n-1} S_i (x - x_i)_+^3 .]$$

Notes, Chapter 4

(1) This theorem is a special case of the *uniform boundedness principle*, but the proof given here may be carried over *in toto* to the more general case. An interesting exposition of the uniform boundedness principle is given in Gál [1].

(2) At this writing, the subject of interpolation and approximation by splines is still in the midst of an extraordinary growth. What we call "splines," i.e., piecewise cubics, are just the simplest instances of much more general objects. As a starting point for further study we recommend the survey by Birkhoff and DeBoor [1], the book by Ahlberg, Nilson, and Walsh [1], and any of the recent work of the father of splines, I. J. Schoenberg [1].

CHAPTER 5

APPROXIMATION AND INTERPOLATION BY RATIONAL FUNCTIONS

Up to this point we have been concerned with approximation by polynomials (or, piecewise polynomial functions). Polynomials have the pleasant property of depending linearly on their coefficients, and so, for example, we were able to rely on the general results of the introductory chapter to establish the existence of best polynomial approximations of given degree. As our final topic we shall study approximation out of a useful family of functions, the members of which do not depend linearly on their parameters. These are the rational functions, that is, functions that are ratios of polynomials.

5.1 Existence, Characterization, and Uniqueness

Let $R(m, n)$ denote the set of rational functions, r, that can be written in the form

$$r(x) = \frac{p(x)}{q(x)},$$

where $p \in P_m$, $q \in P_n$, and r is irreducible, by which we mean that p and q have no zeros in common. Every rational function p/q, $p \in P_m$, $q \in P_n$ is *equivalent* to an irreducible rational function, a member of $R(m, n)$ obtained by cancelling out all zeros that p and q have in common. Thus, for us, all rational functions are irreducible, in the sense that every rational function is replaced by an irreducible equivalent. For example,

$$\frac{x + 1}{x^2 + 3x + 2} = \frac{(x + 1)}{(x + 1)(x + 2)} \tag{5.1.1}$$

is equivalent to

$$\frac{1}{x + 2,} \tag{5.1.2}$$

120

and we always replace (5.1.1) by (5.1.2). We will always also assume that q is not the zero polynomial. Or, if we agree that, in general, a polynomial q has degree n, which we indicate by $\partial q = n$, if $q(x) = b_0 + b_1 x + \cdots + b_n x^n$, $b_n \neq 0$, then nonzero constants have degree 0, and we assign the degree $-\infty$ to the zero polynomial; then we assume always that $\partial q \geq 0$. The zero rational function will always be taken to be $0/1$.

Our first task will be to establish the existence of best approximations out of $R(m, n)$, in the uniform norm, to a given function continuous on an interval $[a, b]$. This is not a consequence of Theorem I.1, since $R(m, n)$ is not a subspace of $C[a, b]$.

THEOREM 5.1.   *If* $f \in C[a, b]$, *there exists* $r^* \in R(m, n)$ *such that*

$$\|f - r^*\| \leq \|f - r\|$$

for all $r \in R(m, n)$, *the norm being the uniform norm.*

Proof.   We first observe that it suffices to restrict our attention to $r = p/q \in R(m, n)$, for which $q \neq 0$ on $[a, b]$, since, if q has a zero on $[a, b]$ (which cannot then be a zero of p since r is assumed to be irreducible), then $\|f - r\| = \infty$, while $r = 0$ is a competitor for best approximation that produces a finite error, $\|f\|$. Indeed, we shall suppose, with no loss in generality, that $r = p/q$ is normalized so that $q > 0$ in $[a, b]$ and $\|q\| = 1$. This normalization can easily be effected by multiplying numerator and denominator of r by a suitable constant.

Let

$$\rho = \inf_{r \in R(m, n)} \|f - r\|,$$

and suppose that

$$\|f - r_i\| = \rho_i, \qquad i = 1, 2, \ldots,$$

with ρ_i monotonically decreasing to ρ, and

$$r_i = \frac{p_i}{q_i} \in R(m, n)$$

$q_i > 0$ on $[a, b]$ and $\|q_i\| = 1$. Then, for i sufficiently large,

$$\|r_i\| \leq \rho + 1 + \|f\| = M, \tag{5.1.3}$$

and, hence,

$$\|p_i\| \leq \|q_i\| \, \|r_i\| \leq M. \tag{5.1.4}$$

Let $x_1, \ldots, x_{m+1}$ be any distinct points in $[a, b]$; then, since $p_i \in P_m$,

$$p_i(x) = \sum_{j=1}^{m+1} p_i(x_j) l_j(x), \tag{5.1.5}$$

the $l_j(x) \in P_m$ being the fundamental polynomials for interpolation at $x_1, \ldots,$ x_{m+1}, which depend only on $x_1, \ldots, x_{m+1}$ and not on p_i (see p. 87). In view of (5.1.4), each sequence of numbers

$$\{p_i(x_j)\}_{i=1}^{\infty}$$

is bounded by M, $j = 1, \ldots, m + 1$. Thus $\{p_i(x_1)\}$ has a convergent subsequence, $p_{i_1}(x_1) \to \bar{p}_1$. Then the sequence $\{p_{i_1}(x_2)\}$ has a convergent subsequence, $p_{i_2}(x_2) \to \bar{p}_2$. Continuing in this fashion for a finite number of steps, we obtain, finally, an infinite sequence of indices $\{i_{m+1}\}$ with the property that

$$p_{i_{m+1}}(x_j) \to \bar{p}_j, \qquad j = 1, \ldots, m + 1.$$

If we rename the sequence $\{i_{m+1}\}$, $\{i\}$, we conclude from (5.1.5) that $p_i(x)$ converges uniformly, as $i \to \infty$, to $\bar{p}(x) \in P_m$. Similarly, after suitable renaming of indices, $q_i(x)$ converges uniformly, as $i \to \infty$, to $\bar{q}(x) \in P_n$, with $\|\bar{q}\| = 1$ and $\bar{q}(x) \geq 0$. If $\bar{q}(x)$ has zeros in $[a, b]$, it has at most n of them, since $\|\bar{q}\| = 1$. At every $x \in [a, b]$, other than one of these zeros, we have

$$|\bar{p}(x)| \leq |\bar{q}(x)| M,$$

and hence, by the continuity of $\bar{p}$ and $\bar{q}$, every zero of $\bar{q}$ in $[a, b]$ must also be a zero of $\bar{p}$. Thus,

$$r^* = \bar{p}/\bar{q} \in R(m, n)$$

satisfies

$$\|f - r^*\| = \rho,$$

and the theorem is proved. ∎

We turn next to a characterization of best rational approximations. It turns out that the characterization resembles that of best polynomials as described in Chapter 1 (Theorem 1.7). Just as in the case of polynomials, a set of N distinct points satisfying $a \leq x_1 < x_2 < \cdots < x_{N-1} < x_N \leq b$ is called an *alternating set* for the error $f - r$ if

$$|f(x_j) - r(x_j)| = \|f - r\|, \qquad j = 1, \ldots, N,$$

and

$$f(x_j) - r(x_j) = -[f(x_{j+1}) - r(x_{j+1})], \qquad j = 1, \ldots, N - 1.$$

We then have

THEOREM 5.2. *If $f \in C[a, b]$, then $r = p/q$ is a best uniform approximation to f out of $R(m, n)$, if and only if $f - r$ has an alternating set consisting of*

$$N = 2 + \max{(n + \partial p, m + \partial q)} \qquad (5.1.6)$$

points.

Proof.   (i) Suppose that $r = p/q \in R(m, n)$ and that $f - r$ has an alternating set, $a \le x_1 < \cdots < x_N \le b$, where N is given by (5.1.6), so that

$$|f(x_j) - r(x_j)| = \|f - r\| = \rho \ge 0, \qquad j = 1, \ldots, N.$$

(Note that ρ must be finite, since $N \ge 2 + \partial q$, and hence q cannot be zero at each x_j without being identically zero, which it cannot be.) If $r_1 \in R(m, n)$ and

$$\rho_1 = \|f - r_1\| < \|f - r\|,$$

then

$$r_1 - r = (f - r) - (f - r_1)$$

alternates in sign as we run through $x_1, \ldots, x_N$ and hence has $N - 1$ zeros in $[a, b]$. But the numerator of $r_1 - r$ belongs to P_k, where

$$k = \max(m + \partial q, n + \partial p) = N - 2.$$

Thus, $r_1 = r$, a contradiction that establishes our result.

(ii) Suppose that $r = p/q \in R(m, n)$ is a best approximation to f, and suppose that $q > 0$ in $[a, b]$, as we may. Let the largest alternating set for $f - r$ consist of $N' < N$ points, $a \le x_1 < x_2 < \cdots < x_{N'} \le b$. Let us put $\|f - r\| = \rho$. (Assume $\rho > 0$; otherwise, the theorem is trivial.) Then, a repetition of the procedure followed in the polynomial case, Theorem 1.7, leads to points $z_1 < z_2 < \cdots < z_{N'-1}$, with the property that $a < z_1$, $z_{N'-1} < b$ and that there exists $\varepsilon > 0$ ($\varepsilon < \rho$) such that, in the intervals $[a, z_1], [z_1, z_2], \ldots, [z_{N'-2}, z_{N'-1}], [z_{N'-1}, b]$, the inequalities

$$-\rho + \varepsilon \le f(x) - r(x) \le \rho \tag{5.1.7}$$

and

$$-\rho \le f(x) - r(x) \le \rho - \varepsilon \tag{5.1.8}$$

hold alternately, and equality does not hold in (5.1.7), (5.1.8) at $x = z_i$, $i = 1, \ldots, N' - 1$. Let

$$k(x) = (x - z_1) \cdots (x - z_{N'-1}).$$

We claim that there exist $p_1 \in P_m$, $q_1 \in P_n$ such that

$$k = qp_1 - q_1p.$$

For suppose that $\partial p \ge \partial q$; then the euclidean algorithm yields

$$\begin{aligned}
p &= u_1q + v_1, & \partial v_1 &< \partial q, \\
q &= u_2v_1 + v_2, & \partial v_2 &< \partial v_1, \\
v_1 &= u_3v_2 + v_3, & \partial v_3 &< \partial v_2, \\
&\;\;\vdots & &\;\;\vdots \\
v_{\mu-1} &= u_{\mu+1}v_\mu + v_{\mu+1}, & \partial v_{\mu+1} &< \partial v_\mu, \\
v_\mu &= u_{\mu+2}v_{\mu+1}.
\end{aligned} \tag{5.1.9}$$

Here, $v_{\mu+1}$ is a factor of p and q, since it is a factor of v_μ, according to the last line of the algorithm. But p/q is irreducible, and hence $v_{\mu+1}$ is a constant. There is no loss in generality in taking this constant to be 1. It is now easy to conclude from (5.1.9) that, since $v_{\mu+1} = 1$ is a linear combination of $v_{\mu-1}$ and v_μ with polynomial coefficients, and, generally, v_k is a linear combination of v_{k-1} and v_{k-2} with polynomial coefficients, there are polynomials α and β such that

$$1 = q\alpha - \beta p, \tag{5.1.10}$$

and that a similar conclusion holds if $\partial q \geq \partial p$. Thus

$$k = q \cdot \alpha k - \beta k p. \tag{5.1.11}$$

If $n + \partial p \geq m + \partial q$, we divide αk by p to obtain

$$\alpha k = \gamma p + \delta,$$

where

$$\partial \delta < \partial p \leq m, \tag{5.1.12}$$

so that (5.1.11) becomes

$$k = q \cdot \delta - (\beta k - \gamma q)p = q \cdot \delta - \lambda p,$$

where, in view of (5.1.12), $\partial(q\delta) < \partial q + \partial p$, hence $\partial(\lambda p) = \partial(k - q\delta) \leq \max(n + \partial p, m + \partial q) = n + \partial p$, so that $\partial \lambda \leq n$.

If $m + \partial q \geq n + \partial p$, we divide βk by q to obtain

$$\beta k = \gamma_1 q + \delta_1,$$

where

$$\partial \delta_1 < \partial q \leq n,$$

so that (5.1.11) becomes

$$k = q(\alpha k - \gamma_1 p) - \delta_1 p = q \cdot \lambda_1 - \delta_1 p,$$

where $\partial(\delta_1 p) < \partial q + \partial p$; hence

$$\partial(q\lambda_1) = \partial(k + \delta_1 p) \leq \max(n + \partial p, m + \partial q) = m + \partial q$$

so that

$$\partial \lambda_1 \leq m.$$

The upshot is that, in either case, there exist polynomials $p_1 \in P_m$ and $q_1 \in P_n$ such that

$$k = qp_1 - q_1 p,$$

as we claimed.

Therefore, if τ is a real number chosen to satisfy

$$|\tau| \, \|q_1\| < \min_{a \le x \le b} q(x),$$

then

$$r_\tau = \frac{p - \tau p_1}{q - \tau q_1}$$

is a member of $R(m, n)$ whose denominator is positive. But

$$f - r_\tau = f - r + \frac{p}{q} - \frac{p - \tau p_1}{q - \tau q_1} = f - r + \tau \frac{k}{q(q - \tau q_1)}, \qquad (5.1.13)$$

and if, further, τ is required to satisfy

$$|\tau| \left\| \frac{k}{q(q - \tau q_1)} \right\| < \frac{\varepsilon}{2},$$

and its sign (which thus far remains at our disposal) is chosen so that $(-1)^{N'-1}\tau$ is positive or negative when (5.1.8) or (5.1.7) respectively hold in $[a, z_1]$, then, in view of (5.1.7) and (5.1.8), we conclude that $\|f - r_\tau\| < \rho$, a contradiction. Thus $N' \ge N$, and the theorem is proved. ∎

As in the case of polynomial approximation, uniqueness now follows.

THEOREM 5.3.   *If $r^* = (p^*/q^*)$ is a best approximation to $f \in C[a, b]$ out of $R(m, n)$, then*

$$\|f - r^*\| < \|f - r\|$$

for all $r \subset R(m, n)$, $r \ne r^*$.

Proof.   Suppose that $r = p/q \in R(m, n)$ and that $\|f - r\| = \|f - r^*\| = \rho$. According to Theorem 5.2, $f - r^*$ has an alternating set, $x_1, \ldots, x_N$, where $N = 2 + \max(n + \partial p^*, m + \partial q^*)$, and $f - r$ has an alternating set, $x'_1, \ldots, x'_{N'}$, where $N' = 2 + \max(n + \partial p, m + \partial q)$. Suppose that $N \ge N'$. Consider

$$s = r - r^* = (f - r^*) - (f - r).$$

If $s(x_i) \ne 0$, then $\operatorname{sgn} s(x_i) = \operatorname{sgn}(f(x_i) - r^*(x_i))$. If $s(x_i) \ne 0$ and $s(x_{i+1}) = s(x_{i+2}) = \cdots = s(x_{i+j-1}) = 0$, while $s(x_{i+j}) \ne 0$, then, since $x_1, \ldots, x_N$ is an alternating set for $f - r^*$, $(-1)^k[f(x_k) - r^*(x_k)]$ is of one sign for $k = 1, \ldots, N$, and so

$$\operatorname{sgn} s(x_i) = (-1)^j \operatorname{sgn} s(x_{i+j}).$$

Therefore, if j is odd, then $s(x)$ has an odd number of zeros (counting multiple zeros according to multiplicity) in $[x_i, x_{i+j}]$, while if j is even, then $s(x)$ has an even number of zeros in $[x_i, x_{i+j}]$. Since, by definition, s has at least $j - 1$ zeros in $[x_i, x_{i+j}]$, we can therefore conclude that s has at least j zeros in $[x_i, x_{i+j}]$. Thus, s has at least $N - 1$ zeros in $[a, b]$. But the zeros of s are zeros of its numerator $p^*q - q^*p$, which is a polynomial of degree at most $\max(\partial p^* + \partial q, \partial q^* + \partial p) \le N' - 2 \le N - 2$. Hence, $s = 0$ and $r = r^*$. ∎

Example. The best approximation to $f(x) = x^2$ on $[-1, 1]$ out of $R(1, 1)$ is $p = \frac{1}{2}$. For $\|x^2 - \frac{1}{2}\| = \frac{1}{2}$ and $f(-1) - \frac{1}{2} = -[f(0) - \frac{1}{2}] = f(1) - \frac{1}{2} = \frac{1}{2}$; thus, $f - \frac{1}{2}$ has an alternating set consisting of $N = 3$ points, while, since $\partial p = \partial q = 0$ and $m = n = 1$, (5.1.6) is satisfied. Of course, $p = \frac{1}{2}$ remains the best approximation out of $R(0, 1)$, $R(1, 0)$, $R(0, 0)$.

If $r^* \in R(m, n)$ is the best approximation to f on $[a, b]$, then $f - r^*$ has an alternating set of N points and, in view of (5.1.6), $N \le m + n + 2$. We call the quantity $d = m + n + 2 - N$ the *deficiency* of f with respect to (m, n). It is clear that

$$d = \min(m - \partial p^*, n - \partial q^*).$$

If $d = 0$, we say that f is *normal* for (m, n); otherwise, we call f *deficient* for (m, n) [or (m, n) *deficient*]. Thus, as the immediately preceding example shows, $f = x^2$ is deficient for $(1, 1)$ but normal for $(0, 1)$, $(1, 0)$, and $(0, 0)$. It is the phenomenon of deficiency that makes the study of rational approximation more complicated than that of polynomial approximation.

It is clear that if r^* is the best approximation to f out of $R(m, n)$, it is also the best approximation to f out of $R(k, l)$ for $\partial p^* \le k \le m$ and $\partial q^* \le l \le n$. Also, if f has (m, n) deficiency d, then $r^* \in R(m - d, n - d)$ but

$$r^* \notin R(m - d - 1, n - d - 1).$$

If f is normal for (m, n), $m, n = 0, 1, 2, \ldots$, we call f *hypernormal*.

THEOREM 5.4. *If f is hypernormal and $r^* = p^*/q^*$ is the best approximation to f out of $R(m, n)$, then $\partial p^* = m$ and $\partial q^* = n$. Moreover, the largest alternating set for $f - r^*$ consists of $m + n + 2$ points.*

Proof. Suppose that, say, $\partial p^* \le m - 1$, and hence, since f is normal for (m, n), $\partial q^* = n$. Then r^* is the best approximation to f out of $R(m, n + 1)$, since $f - r^*$ has an alternating set consisting of

$$2 + \max(n + \partial p^*, m + \partial q^*)$$
$$= n + m + 2 = 2 + \max(n + 1 + \partial p^*, m + \partial q^*)$$

points. Hence f has $(m, n + 1)$ deficiency 1; therefore it is not hypernormal. A similar proof holds if $\partial q^* \le n - 1$.

To prove the final statement of the theorem, suppose it to be false; i.e., suppose that $f - r^*$ has an alternating set consisting of $m + n + 3$ points. Of course, we know $f - r^*$ has an alternating set made up of $m + n + 2$ points. The point of this result is that $f - r^*$ has no larger alternating set (cf. Exercise 5.3). But since, as we have just finished proving, $\partial p^* = m$, $\partial q^* = n$,

$$m + n + 3 = 2 + \max{(n + 1 + \partial p^*, m + 1 + \partial q^*)},$$

hence r^* is the best approximation to f out of $R(m + 1, n + 1)$. Thus f has $(m + 1, n + 1)$ deficiency 1, contradicting its hypernormality. ∎

Thus hypernormality implies "full" alternating sets, which makes the rational approximation problem polynomial-like. It seems worthwhile, therefore, to have some conditions that imply hypernormality.

THEOREM 5.5.   *If nonrational $f \in C[a, b]$ is not zero in $[a, b]$, and if, for each (k, l), every function that is a linear combination of $1, x, \ldots, x^k, f(x)$, $xf(x), \ldots, x^l f(x)$ has at most $k + l + 1$ distinct zeros in $[a, b]$, except the zero function, then f is hypernormal.*

Proof.   Suppose that the theorem is false and that f is deficient for (m, n), with deficiency $d \geq 1$. Let $r^* = p^*/q^*$ be the best approximation to f out of $R(m, n)$; then $f - r^*$ has an alternating set of $n + m + 2 - d$ points and so has $n + m + 1 - d$ (distinct) zeros in $[a, b]$. If $p^* = 0$, then f has a zero in $[a, b]$, contrary to our hypothesis. If $p^* \neq 0$, $fq^* - p^*$ has $n + m + 1 - d = (n - \partial q^*) + (m - \partial p^*) - d + \partial q^* + \partial p^* + 1$ zeros in $[a, b]$, and since $d = \min{(n - \partial q^*, m - \partial p^*)} \geq 1$, $fq^* - p^*$ has more than $\partial q^* + \partial p^* + 1$ zeros in $[a, b]$, which implies (take $k = \partial p^*, l = \partial q^*$) that $f = p^*/q^*$, a rational function. This contradiction proves the theorem. ∎

Example. e^x is hypernormal on $[a, b]$. For, certainly $e^x \neq 0$ in $[a, b]$. Suppose that $a_0 + a_1 x + \cdots + a_k x^k + b_0 e^x + b_1 x e^x + \cdots + b_l x^l e^x = p(x) + e^x q(x)$ has $k + l + 2$ distinct zeros in $[a, b]$ and that $b_l \neq 0$. If we apply Rolle's Theorem $k + 1$ times to $p(x) + e^x q(x)$, we can conclude that

$$h(x) = \frac{d^{k+1}[e^x q(x)]}{dx^{k+1}}$$

has $l + 1$ zeros in $[a, b]$. But an application of Leibniz's rule (Courant [1], Vol. I, p. 202) yields

$$h = e^x \left[q + \binom{k + 1}{1} q' + \binom{k + 1}{2} q'' + \cdots + \binom{k + 1}{k} q^{(k)} + q^{(k+1)} \right].$$

The quantity in square brackets is a polynomial of degree l having $l + 1$ zeros, and hence is identically zero. This implies that the coefficient of x^l of this polynomial, which is b_l, must be zero, contrary to our assumption. The hypernormality of e^x now follows from Theorem 5.5.

5.2 Degree of Approximation

We turn now to the question of how closely it is possible to approximate a given function by rational functions. The theory is far less well developed than in the polynomial case.

Let us put

$$E_{m,n} = E_{m,n}(f; [a, b]) = \| f - r^* \|,$$

where r^* is the best approximation to f out of $R(m, n)$. Then, if $k, l \geq 0$, $E_{m+k,n+l} \leq E_{m,n}$ and, in particular, $E_{m,n} \leq E_{m,0}$ for $n \geq 0$. Therefore, if $f \in C[a, b]$,

$$\| f - r^* \| \leq E_{m,0}(f; [a, b]) \leq 6\omega\left(\frac{b - a}{2m}\right), \tag{5.2.1}$$

in view of Corollary 1.4.1. It may happen that the bound in (5.2.1) can be substantially improved. This fact is illustrated by the following remarkable example due to D. J. Newman [1].

THEOREM 5.6.   *If $n > 4$ and $f(x) = |x|$, then, if n is even,*

$$E_{n,n}(f; [-1, 1]) \leq 3e^{-\sqrt{n}},$$

and, if n is odd,

$$E_{n+1,n-1}(f; [-1, 1]) \leq 3e^{-\sqrt{n}}.$$

Proof.   Put

$$t = e^{-1/\sqrt{n}}, \qquad p(x) = \prod_{j=0}^{n-1} (x + t^j), \qquad \text{and} \qquad r(x) = \frac{x[p(x) - p(-x)]}{p(x) + p(-x)}.$$

Then $r(x) \in R(n, n)$ if n is even, $r(x) \in R(n + 1, n - 1)$ if n is odd, and $r(x)$ is an even function for each n.

If $x \geq 0$, it is obvious that $e^x \leq 1 + x$. Also, if $h(x) = 1 - x - e^{-x}$, then $h(0) = 0$, $h'(x) \leq 0$ for $x \geq 0$, and so $1 - x \leq e^{-x}$ for $x \geq 0$. Hence

$$\frac{1 - x}{1 + x} \leq e^{-x} \cdot e^{-x} = e^{-2x}, \qquad x \geq 0,$$

and so

$$\prod_{j=1}^{n} \frac{1 - t^j}{1 + t^j} \leq \exp\left(-2 \sum_{j=1}^{n} t^j\right) = \exp\left(-2t \frac{1 - t^n}{1 - t}\right). \tag{5.2.2}$$

Now, since $n \geq 5$,

$$2t(1 - t^n) = 2e^{-1/\sqrt{n}}(1 - e^{-\sqrt{n}}) \geq 2e^{-1/\sqrt{5}}(1 - e^{-\sqrt{5}}) \geq 1,$$

and, as we have seen, $1 - x \leq e^{-x}$, $x \geq 0$; hence, if we take $x = 1/\sqrt{n}$, we obtain

$$1 - \frac{1}{\sqrt{n}} \leq e^{-1/\sqrt{n}} \quad \text{or} \quad \frac{1}{1 - t} = \frac{1}{1 - e^{-1/\sqrt{n}}} \geq \sqrt{n}.$$

Thus, (5.2.2) implies that

$$\prod_{j=1}^{n} \frac{1 - t^j}{1 + t^j} \leq e^{-\sqrt{n}}. \tag{5.2.3}$$

Suppose that $0 \leq x \leq e^{-\sqrt{n}}$. Then $p(-x) > 0, p(x) > 0$, and $p(-x) \leq p(x)$ so that

$$0 \leq r(x) \leq x,$$

and hence

$$||x| - r(x)| = x - r(x) \leq x \leq e^{-\sqrt{n}}. \tag{5.2.4}$$

Suppose that $e^{-\sqrt{n}} \leq x \leq 1$. If $t^{j+1} \leq x \leq t^j$, $0 \leq j \leq n - 1$, then, in view of (5.2.3),

$$\left| \frac{p(-x)}{p(x)} \right| = \prod_{k=0}^{j} \frac{t^k - x}{t^k + x} \prod_{k=j+1}^{n-1} \frac{x - t^k}{x + t^k} \leq \prod_{k=0}^{j} \frac{t^k - t^n}{t^k + t^n} \prod_{k=j+1}^{n-1} \frac{t^j - t^k}{t^j + t^k}$$

$$= \prod_{m=1}^{n} \frac{1 - t^m}{1 + t^m} \leq e^{-\sqrt{n}}. \tag{5.2.5}$$

Thus

$$||x| - r(x)| = |x - r(x)| = 2x \left| \frac{p(-x)}{p(x) + p(-x)} \right| = \frac{2x}{|1 + [p(x)/p(-x)]|}$$

$$\leq \frac{2}{|[p(x)/p(-x)] - 1|}.$$

But, according to (5.2.5),

$$\left| \frac{p(x)}{p(-x)} \right| \geq e^{\sqrt{n}},$$

so that

$$||x| - r(x)| \leq \frac{2}{e^{\sqrt{n}} - 1} \leq 3e^{-\sqrt{n}}, \tag{5.2.6}$$

since $n \geq 5$. (5.2.4) and (5.2.6) together yield

$$||x| - r(x)| \leq 3e^{-\sqrt{n}}, \qquad 0 \leq x \leq 1,$$

and, since $|x|$ and $r(x)$ are both even functions,

$$||x| - r(x)| \leq 3e^{-\sqrt{n}}, \quad -1 \leq x \leq 1.$$

The theorem now follows from the observation that the best approximations out of $R(n, n)$ (n even), $R(n + 1, n - 1)$ (n odd), produce errors no larger than that produced by $r(x)$. ∎

Newman [1] also shows that, if $r \in R(l, k)$, $l, k \leq n$, then

$$\max_{-1 \leq x \leq 1} ||x| - r(x)| > \frac{e^{-9\sqrt{n}}}{2},$$

and so, in this sense, the rational functions constructed in Theorem 5.6 are not too far from best approximations. If we recall (Exercise 2.26) that, in the polynomial case

$$E_n(|x|; [-1, 1]) > \frac{c}{n},$$

the significance of Theorem 5.6 becomes clear. Rational functions seem capable of dramatically better approximative power than polynomials, for some functions at least. The theory of error due to rational approximation has not yet been fully developed, and we can say no more about it here. Other results along the Newman line (Theorem 5.6) have recently been obtained by Turan [1]. See, also, Lorentz [1, p. 82].

5.3   Finite Point Sets

When we turn to rational approximation on a finite set of points, some new problems arise. In the first place, there may not be a best approximation. For example, let us try to approximate the values $f(0) = 1$, $f(1) = 0$ out of $R(0, 1)$. The functions

$$r(x) = \frac{a}{bx + a}$$

are in $R(0, 1)$ so long as $|a| + |b| > 0$. They yield $r(0) = 1$ and $r(1) = a/(b + a)$; hence, given $\varepsilon > 0$, by choosing b large enough, we have

$$\max \{|f(0) - r(0)|, |f(1) - r(1)|\} = |f(1) - r(1)| < \varepsilon.$$

The best approximation out of $R(0, 1)$ would have to produce an error of zero, which can only happen if there is a function $a/(bx + c)$, having the value 0 at $x = 1$; that is, if $a = 0$. But then the function cannot be 1 at $x = 0$. There is no best approximation.

Secondly, the direct analogue of Theorem 5.2 may fail. For example, consider the values $f(-1) = -1, f(\frac{1}{2}) = 2, f(1) = 1$. The rational function

$$r(x) = \frac{-9}{16x - 20}$$

is a member of $R(0, 1)$ and satisfies

$$-[f(-1) - r(-1)] = [f(\tfrac{1}{2}) - r(\tfrac{1}{2})] = -[f(1) - r(1)] = \tfrac{5}{4},$$

so that the points $-1, \frac{1}{2}, 1$ are an alternating set for $f - r$ on the set $\{-1, \frac{1}{2}, 1\}$ on which we are approximating, yet $1/x \in R(0, 1)$ is clearly a better approximation than r. The trouble stems from the fact that $1/x$ is not continuous in the smallest interval containing our discrete point set.

A correct analogue of Theorem 5.2, which provides us with a characterization of best approximations on a finite point set, is easily obtained. Suppose that $X_m: \{x_1, \ldots, x_m\}$ satisfies $a = x_1 < x_2 < \cdots < x_{m-1} < x_m = b$. Let

$$\bar{R}(m, n) = R(m, n) \cap C([a, b]),$$

that is, $\bar{R}(m, n)$ consists of rational functions $r = p/q$ such that $p \subset P_m$, $q \in P_n$, and q has no zeros on $[a, b]$. Then it is easy to obtain, by following the proof of Theorem 5.2,

THEOREM 5.7.   *If f is defined on X_m, then $r = p/q$ is a best uniform approximation to f (on X_m) out of $\bar{R}(m, n)$, if and only if $f - r$ has an alternating set consisting of*

$$N = 2 + \max{(n + \partial p, m + \partial q)}$$

points of X_m.

The uniqueness of best approximations out of $\bar{R}(m, n)$ now follows.

Theorem 5.7 by no means answers the existence question, but it does provide us with a criterion for existence. The delicacy of the existence question is typified by the following special result (for proof of this result see the original paper).

THEOREM 5.8 (WERNER [4]).   *f has a best approximation out of $\bar{R}(1, 1)$ on X_4 if*

$$\text{sgn} [f(x_1) - f(x_3)] = \text{sgn} [f(x_2) - f(x_4)]. \qquad (5.3.1)$$

If there exists $r \in \bar{R}(1, 1)$ satisfying

$$f(x_i) - r(x_i) = (-1)^{i+1}(f(x_1) - r(x_1)), \qquad i = 2, 3, 4,$$

then (5.3.1) *holds.*

5.4 Rational Interpolation

The task of rational interpolation is to determine a rational function $r \in R(m, n)$ that satisfies

$$r(x_i) = f_i, \qquad i = 1, \ldots, k, \tag{5.4.1}$$

where $x_1, \ldots, x_k$ are distinct real numbers, $f_1, \ldots, f_k$ are arbitrary values, and $k = m + n + 1$. $m + n + 1$ is the largest value of k for which one could hope to satisfy (5.4.1) in general, since there are precisely $m + n + 1$ parameters at our disposal in fixing r (we can always divide through by one of the nonzero coefficients). It may be possible for some sets (x_i, f_i), $i = 1, \ldots, k$, to satisfy (5.4.1) by means of an $r \in R(\mu, \nu)$ with $\mu + \nu < k - 1$, and $\mu \le m$, $\nu \le n$. Such sets (x_i, f_i) we call *degenerate* configurations.

In the case of polynomial interpolation (the special case $R(m, 0)$ of the present discussion), as we saw in Chapter 4, (5.4.1) can always be satisfied. Such is not the case for rational interpolation in general. For example, it is clear that, if some of the f_i are zero, and if $m = 0$, (5.4.1) cannot be satisfied. A more organic difficulty appears in the case $m = n = 1$. Suppose that $f_1 = f_2 \ne f_3$. (Here, $k = 3$.) We require that

$$r(x) = \frac{a_0 + a_1 x}{b_0 + b_1 x} \tag{5.4.2}$$

satisfy

$$f_1 = \frac{a_0 + a_1 x_1}{b_0 + b_1 x_1} = \frac{a_0 + a_1 x_2}{b_0 + b_1 x_2} = f_2.$$

Thus

$$a_0 b_1 (x_2 - x_1) - a_1 b_0 (x_2 - x_1) = 0,$$

or $a_0 b_1 = a_1 b_0$. Suppose that $b_1 \ne 0$. If we put $a_0 = (a_1 b_0)/b_1$ in (5.4.2), we see that r is a constant. If $b_1 = 0$, then $a_1 = 0$, since not both b_0 and b_1 are zero, and again r is constant. Thus $r = f_1$, but then $r(x_3) \ne f_3$. The interpolation problem cannot be solved for this set of data.

In order to determine when the interpolation problem (5.4.1) has a solution and how to find the solution, we need to examine the problem more carefully. We begin with a definition.

DEFINITION. $r_1 = p_1/q_1$ and $r_2 = p_2/q_2 \in R(m, n)$ are equal if

$$p_1 q_2 = q_1 p_2.$$

THEOREM 5.9. *The interpolation problem in $R(m, n)$, that is, the problem of satisfying (5.4.1), has at most one solution.*

Proof. If $r_1, r_2 \in R(m, n)$ satisfy (5.4.1), then $r_1(x_i) - r_2(x_i) = 0$, $i = 1, \ldots, m + n + 1$, and, hence, $p_1(x_i)q_2(x_i) - p_2(x_i)q_1(x_i) = 0$. But $p_1q_2 - p_2q_1 \in P_{m+n}$ and is therefore zero. ∎

If $r = p/q$ satisfies (5.4.1), the coefficients of p and q satisfy the following system of homogeneous linear equations:

$$q(x_i)f_i - p(x_i) = (b_0 + b_1x_i + \cdots + b_nx_i^n)f_i$$
$$- (a_0 + \cdots + a_mx_i^m) = 0, \qquad i = 1, \ldots, k. \quad (5.4.3)$$

(5.4.3) consists of k equations in $k + 1$ unknowns and therefore always has a nontrivial solution. Moreover, each nontrivial solution of (5.4.3) defines a bona fide rational function; that is, no nontrivial solution has $b_0 = b_1 = \cdots = b_n = 0$. For if this were the case, then we would have $p(x_i) = 0$, $i = 1, \ldots, k$, hence $p = 0$, that is, $a_0 = \cdots = a_m = 0$. Indeed, we have

THEOREM 5.10. *All nontrivial solutions of* (5.4.3) *define the same rational function.*

Proof. Suppose that $q_1, q_2 \in P_n$, $p_1, p_2 \in P_m$ and

$$q_1(x_i)f_i - p_1(x_i) = q_2(x_i)f_i - p_2(x_i) = 0, \qquad i = 1, \ldots, k.$$

Then

$$p_1(x_i)q_2(x_i) - q_1(x_i)p_2(x_i) = 0, \qquad i = 1, \ldots, k,$$

and $p_1q_2 - q_1p_2 \in P_{k-1}$. ∎

Thus the linear equations (5.4.3) lead to a unique rational function, call it $\bar{r}, \in R(m, n)$. If (5.4.1) is satisfied, then $\bar{r}$ is the rational function that interpolates. However, it may happen that $\bar{r}$ does not satisfy (5.4.1) for some i. We call the set of points (x_i, f_i) for which

$$\bar{r}(x_i) \neq f_i \qquad (5.4.4)$$

unattainable points. If $\bar{r} = \bar{p}/\bar{q}$ and (5.4.4) holds, then, since, by definition,

$$\bar{p}(x_i) = f_i\bar{q}(x_i),$$

we must have $\bar{q}(x_i) = \bar{p}(x_i) = 0$; that is, x_i is a common zero of $\bar{p}$ and $\bar{q}$. Roughly speaking, the unattainable points lead to cancellations in the numerator and denominator of the rational function defined by the solution of the linear system (5.4.3).

Example. Let us try to find the $r \in R(1, 1)$ that passes through $(-1, 1)$, $(0, 0)$, $(1, 1)$. We set up equations (5.4.3)

$$b_0 - b_1 - a_0 + a_1 = 0,$$
$$-a_0 = 0,$$
$$b_0 + b_1 - a_0 - a_1 = 0,$$

that have a nontrivial solution $a_0 = 0, a_1 = 1, b_0 = 0, b_1 = 1$, which leads to

$$\bar{r} = \frac{x}{x} = 1.$$

The point $(0, 0)$ is unattainable.

Note that, in this example, the "interpolable" points form a degenerate configuration. This is true in general. Let us call the points that are not unattainable *attainable* points.

THEOREM 5.11. *The attainable points among* (x_i, f_i), $i = 1, \ldots, k$, *form a degenerate configuration, if there exist unattainable points.*

Proof. Let h be the number of unattainable points. Corresponding to each is a common zero of $\bar{p}, \bar{q}$. Upon cancelling out these common zeros, we have $\bar{r} \in R(m - h, n - h)$ and $\bar{r}(x_i) = f_i$ for the $k - h$ attainable points. Since $m + n - 2h < m + n - h = k - h - 1$, the attainable points are degenerate. ∎

As a partial converse to Theorem 5.11 we have

THEOREM 5.12. *If* (x_i, f_i), $i = 1, \ldots, k$, *is a degenerate configuration, that is, there exists* $r_1 \in R(m - s, n - t)$, $s, t \geq 0$, $s + t > 0$, *such that* $r_1(x_i) = f_i$, $i = 1, \ldots, k$; *then for arbitrary* $x_{k+1} \neq x_i$ *and* f_{k+1} *either* $r_1(x_{k+1}) = f_{k+1}$ *or* (x_{k+1}, f_{k+1}) *is unattainable in the set* (x_i, f_i), $i = 1, \ldots, k + 1$, *for* $R(m + t, n + s)$.

Proof. Suppose $r \in R(m + t, n + s)$, $r(x_{k+1}) = f_{k+1}$ but $r(x_i) \neq f_i$ for exactly h other indices. Then $r - r_1$ has $k - h$ zeros, and $pq_1 - qp_1 \in P_l$ where $l = m + n - h < k - h$. Thus, $r = r_1$ and $r_1(x_{k+1}) = f_{k+1}$. ∎

It is now clear that the interpolation problem has a solution if the denominator of the rational function determined by the solution to (5.4.3) does not have any of the interpolating nodes $x_1, \ldots, x_k$ as a zero. Our interest thus turns to a study of solutions of (5.4.3). To this end, we employ an ingenious device that goes back to Jacobi.

Let $\omega(x) = (x - x_1) \cdots (x - x_k)$. Then, in view of p. 35,

$$\sum_{i=1}^{k} \frac{g_i}{\omega'(x_i)}$$

is the leading coefficient of the polynomial of degree at most $k - 1$ that takes the value g_i at x_i, $i = 1, \ldots, k$. Any solution of (5.4.3) is also a solution of

$$\frac{x_i^s p(x_i)}{\omega'(x_i)} = \frac{x_i^s f_i q(x_i)}{\omega'(x_i)}, \qquad i = 1, \ldots, k, \qquad s = 0, \ldots, n - 1,$$

and hence of

$$\sum_{i=1}^{k} \frac{x_i^s p(x_i)}{\omega'(x_i)} = \sum_{i=1}^{k} \frac{x_i^s f_i q(x_i)}{\omega'(x_i)}, \qquad s = 0, \ldots, n-1. \tag{5.4.5}$$

But since $x^s p \in P_{k-2}$, its leading coefficient (the coefficient of x^{k-1}) is zero, and so the left-hand side of (5.4.5) is zero. Thus, the coefficients of q, $(b_0, \ldots, b_n)$ satisfy

$$\sum_{i=1}^{k} \frac{x_i^s f_i q(x_i)}{\omega'(x_i)} = \sum_{j=0}^{n} v_{sj} b_j = 0, \qquad s = 0, \ldots, n-1, \tag{5.4.6}$$

where

$$v_{sj} = \sum_{i=1}^{k} \frac{x_i^{s+j} f_i}{\omega'(x_i)}, \qquad j = 0, \ldots, n, \qquad s = 0, \ldots, n-1. \tag{5.4.7}$$

Note that the coefficients v_{sj} depend only on the sum $s + j$. Once $b_0, \ldots, b_n$ have been determined from (5.4.6), we can ascertain whether $q(x_i) = 0$, $i = 1, \ldots, k$, or not. If no x_i is a zero of q, we know that the interpolation problem has a solution, and it is given by $r = p/q$, where p is now determined by

$$p(x_i) = f_i q(x_i), \qquad i = 1, \ldots, k.$$

Since $f_i q(x_i)$ are, at this point, known quantities, p may be determined from the Lagrange interpolating formula for polynomials, namely,

$$p(x) = \omega(x) \sum_{i=1}^{k} \frac{p(x_i)}{\omega'(x_i)(x - x_i)} = \omega(x) \sum_{i=1}^{k} \frac{f_i q(x_i)}{\omega'(x_i)(x - x_i)}. \tag{5.4.8}$$

Example. Let us interpolate $|x|$ at $x = -1, -\frac{1}{2}, 0, \frac{1}{2}, 1$ by a function in $R(2, 2)$. We obtain $v_{00} = -\frac{4}{3}$, $v_{10} = v_{01} = 0$, $v_{11} = v_{02} = \frac{2}{3}$, $v_{12} = 0$, and conclude that

$$r = \frac{3x^2}{1 + 2x^2}.$$

However, if we attempt to interpolate the same data by means of a function in $R(3, 1)$, we find that $q(x) = x$, so that $(0, 0)$ is unattainable. The resulting function is $r = \frac{1}{3}(1 + 2x^2)$, so that the points $(-1, 1)$, $(-\frac{1}{2}, \frac{1}{2})$, $(\frac{1}{2}, \frac{1}{2})$, $(1, 1)$ are a degenerate configuration, as expected.

5.5   Computing a Best Approximation

There are many difficulties in actually trying to produce a best rational approximation to a given function, and this topic is still being studied vigorously. We shall content ourselves with sketching one approach and refer the reader to the growing literature for the details.

Suppose that $f(x)$ is continuous on I: $[-1, 1]$ and *normal* for (m, n). The procedure we are about to describe aims at finding an $r \in R(m, n)$ and an alternating set for $f - r$ consisting of $N = m + n + 2$ points. Thus, the procedure is founded on Theorem 5.2. The following lemma, whose analogue in the polynomial case (Theorem 1.13) was an important ingredient in the strategy of the Exchange Method (see p. 40), also helps determine the closely related strategy here. We introduce nomenclature similar to that used in the polynomial case. A set, X, of $N = m + n + 2$ distinct points of I is called a *reference*, and the points of this set are denoted by x_σ. Greek subscripts run over the indices $1, \ldots, N$, and we assume that $x_1 < x_2 < \cdots < x_N$.

LEMMA 5.1. *If X is a reference and $\bar{r}$ a best approximation to f out of $\bar{R}(m, n)$ on X (that is, $\bar{r}$ is continuous on $[x_1, x_N]$), then*

$$\|f - \bar{r}\|_X \le \|f - r\|_{X^*} = \|f - r\| = E_{m,n}(f; I), \qquad (5.5.1)$$

where r is the best approximation to f on I and X^* is an alternating reference for $f - r$. Equality holds in (5.5.1) only if $\bar{r} = r$.

Proof.

$$\|f - \bar{r}\|_X \le \|f - r\|_X \le \|f - r\| = \|f - r\|_{X^*}.$$

The last sentence follows from the uniqueness of best approximations. ∎

The algorithm, sometimes called the Remez algorithm, proceeds as follows:

(i) Let X_0 be an initial reference. Find the best approximation to f on X_0 out of $\bar{R}(m, n)$. Call it r_0. If r_0 exists and f is normal for (m, n) on X_0, then $r_0 = p_0/q_0$ can be determined in the following manner. The N homogeneous linear equations in N unknowns (the unknowns are the coefficients of p_0 and q_0),

$$[f(x_\sigma) - (-1)^\sigma A_0]q_0(x_\sigma) - p_0(x_\sigma) = 0, \qquad (5.5.2)$$

have a nontrivial solution if, and only if, the $N \times N$ determinant of its coefficients, whose kth row is

$$[(f(x_k) - (-1)^k A_0), (f(x_k) - (-1)^k A_0)x_k, \ldots,$$
$$(f(x_k) - (-1)^k A_0)x_k^n, -1, -x_k, \ldots, -x_k^m],$$

is zero. This requires the deviation A_0 to satisfy a polynomial equation of degree at most N. Call this equation $\phi_0(x) = 0$. It can be shown (Werner [1]) that all the zeros of ϕ_0 are real. When one of these zeros is substituted for A_0 in (5.5.2), the resulting linear system can be solved for q_0 and p_0 (most expeditiously by using the technique ascribed to Jacobi on p. 134). It turns out that *at most one* zero of ϕ_0 leads, in this manner, to an $r_0 \in \bar{R}(m, n)$, i.e., to a q_0 with no zero in the interval $[x_1, x_N]$. Under our assumptions that r_0 exists and f is (m, n) normal on X_0, the procedure just described finds r_0, uniquely.

(ii) We now have at hand $r_0 \in \bar{R}(m, n)$ and A_0 satisfying

$$f(x_\sigma) - r_0(x_\sigma) = (-1)^\sigma A_0$$

for $x_\sigma \in X_0$. Suppose, for the sake of concreteness, that $A_0 > 0$. A new reference, X_1, is now chosen. Surround each $x_\sigma \in X_0$ by a closed interval, I_0^σ (in case x_σ is an end point of I, let x_σ be an end point of I_0^σ), so that consecutive intervals I_0^σ and $I_0^{\sigma+1}$ have only one point in common and so that

$$I = \bigcup_{\sigma=1}^{N} I_0^\sigma.$$

Let

$$B_\sigma = \max_{x \in I_0^\sigma} [f(x) - r_0(x)] = [f(x_\sigma^{(1)}) - r_0(x_\sigma^{(1)})], \qquad \sigma \text{ even,}$$

$$B_\sigma = \min_{x \in I_0^\sigma} [f(x) - r_0(x)] = [f(x_\sigma^{(1)}) - r_0(x_\sigma^{(1)})], \qquad \sigma \text{ odd,}$$

any choice of the $x_\sigma^{(1)}$ consonant with their definition being permitted. Clearly, for every σ, $|B_\sigma| \geq A_0$. Let

$$B_0 = \max_\sigma |B_\sigma|;$$

then, from Lemma 5.1 and the definition of $E_{m,n}$, we have

$$A_0 \leq E_{m,n}(f; I) \leq B_0.$$

If $A_0 = B_0$, our quest is over, and r_0 is the best approximation. If $A_0 < B_0$, we return to (i) with X_0 replaced by

$$X_1 = \{x_1^{(1)}, \ldots, x_N^{(1)}\}$$

and A_0 by A_1, and repeat the procedure. The difference $B_j - A_j$ can be inspected to indicate when the process should be stopped.

The assumptions that r_0 exists and that f is normal on X_0 are large ones. It can be shown, however, that if f is normal for (m, n) on I and the initial reference is near enough to an alternating set (or if the initial reference is derived from a rational function r such that $\|f - r\|$ is near enough to $E_{m,n}$), then the assumptions are satisfied and the Remez procedure converges (see Ralston [1] and Werner [3]). In the absence of a good initial guess, experience suggests the zeros of the Chebyshev polynomial, $T_N(x)$, as a good choice for

X_0. Alternate methods of finding best rational approximations are discussed in Cheney and Southard [1].

Exercises

5.1   Find the best approximation out of $R(1, 1)$ to $f(x) = x^3$ on $[-1, 1]$.

5.2   Find the best approximations out of $R(m, n)$ for all m, n to $f(x) = 2x^2 - 1$ on $[-1, 1]$.

5.3   Suppose that $r^* = p^*/q^*$ is the best approximation out of $R(m, n)$ to f and that $f - r^*$ has an alternating set consisting of $N' > N$ points; then show that r^* is the best approximation to f out of $R(m + m', n + n')$, where m', n' are the largest integers for which

$$N' = 2 + \max (n + n' + \partial p^*, m + m' + \partial q^*).$$

5.4   If f is hypernormal, show that $r = 0$ is not the best approximation to f on $[a, b]$ for any (m, n).

5.5   Can a function be deficient for $(0, 0)$?

5.6   Show that, if f is continuous on $[a, b]$ and r^* is its best approximation out of $R(m, n)$, then $f - r^*$ cannot have an alternating set consisting of infinitely many points of $[a, b]$, unless $f = r^*$.

[*Hint:* Show, more generally, that a continuous function g on $[a, b]$ cannot alternately have the value $\pm M$ $(M > 0)$ at infinitely many points of $[a, b]$. (What value would g have at a limit point of such points?)]

5.7   (Boehm [1])   Show that the best approximation out of $R(m, n)$ to f is a polynomial for all (m, n) if, and only if, f is a constant.

[*Hint:* If f is not a constant and r_0^* is its best approximation out of $R(0, 0)$, the largest alternating set of $f - r^*$ consists of k points, in view of Exercise 5.6. Can the best approximation to f out of $R(0, k)$ then be a polynomial?]

5.8   Show that f is (m, n) normal if, and only if, $E_{m,n}(f) < E_{m-1,n-1}(f)$ $(m, n \geq 1)$.

5.9   If, in the hypothesis of Theorem 5.5, the phrase "for each (k, l)" is replaced by "for each (k, l), $k \leq m, l \leq n$," show that f is normal for (m, n).

5.10 $f = x^2$ is deficient for $R(1, 1)$ on $[-1, 1]$. What if the interval is modified slightly to $[-1 + \varepsilon, 1]$?

5.11   If f is (m, n) normal and

$$\|f - g\| < \frac{E_{m-1,n-1}(f) - E_{m,n}(f)}{2},$$

then show that g is (m, n) normal.

5.12   Show that the matrix of coefficients of the linear equations (5.4.6) has the form

$$\begin{pmatrix} v_0 & v_1 & \cdots & v_n \\ v_1 & v_2 & \cdots & v_{n+1} \\ \vdots & & & \vdots \\ v_{n-1} & v_n & \cdots & v_{2n-1} \end{pmatrix},$$

where

$$v_j = \sum_{i=1}^{k} \frac{x_i^j f_i}{\omega'(x_i)}, \qquad j = 0, \ldots, 2n - 1,$$

and $\omega(x) = (x - x_1) \cdots (x - x_k)$.

5.13   Suppose that

$$w_j(x) = \sum_{i=1}^{k} \frac{x_i^j f_i}{\omega'(x_i)(x - x_i)}.$$

Show that

$$w_{j+1}(x) = x w_j(x) - v_j, \qquad j = 0, \ldots, n - 1.$$

(v_j is as defined in Exercise 5.12).

5.14   If $q(x) = b_0 + b_1 x + \cdots + b_n x^n$, show that

$$p(x) = \omega(x)[b_0 w_0(x) + b_1 w_1(x) + \cdots + b_n w_n(x)].$$

5.15   Show that the following determinants give $r = p/q$:

$$q(x) = \begin{vmatrix} 1 & x & x^2 & \cdots & x^n \\ v_0 & v_1 & \cdots & & v_n \\ v_1 & v_2 & \cdots & & v_{n+1} \\ \vdots & & & & \vdots \\ v_{n-1} & \cdots & & & v_{2n-1} \end{vmatrix},$$

$$p(x) = \omega(x) \begin{vmatrix} w_0(x) & w_1(x) & \cdots & w_n(x) \\ v_0 & v_1 & \cdots & v_n \\ v_1 & v_2 & \cdots & v_{n+1} \\ \vdots & & & \vdots \\ v_{n-1} & v_n & \cdots & v_{2n-1} \end{vmatrix}.$$

5.16   If f_1, f_2, f_3 are distinct, show that the $r \in R(1, 1)$ that interpolates satisfies

$$\frac{r - f_1}{r - f_3} = \frac{x - x_1}{x - x_3} \cdot \frac{x_2 - x_3}{x_2 - x_1} \cdot \frac{f_2 - f_1}{f_2 - f_3}.$$

5.17   Show that

$$r(x) = \frac{1 + \lambda x}{1 - \lambda x}$$

agrees with e^x at $x = -1, 0, 1$, where $\lambda = \tanh \frac{1}{2}$, which is approximately 0.462. Obtain an upper bound for the error $|r(x) - e^x|$ on the interval $[-1, 1]$. [*Hint:* Use the power series expansions about $x = 0$ of e^x and $r(x)$, or the result in the next problem.]

5.18   Show that, if $r = p/q \in R(m, n)$ interpolates $f(x)$ at $x_1 < x_2 < \cdots < x_k$, where $k = m + n + 1$ and $f \in C^k[x_1, x_k]$, then, for each $x \in [x_1, x_k]$,

$$f(x) - \frac{p(x)}{q(x)} = \frac{(x - x_1)\cdots(x - x_k)}{k!\,q(x)} [f(\xi)q(\xi)]^{(k)},$$

where the kth derivative of fq is evaluated at some $\xi(x)$ in $[x_1, x_k]$ (cf. Exercise 4.2).

5.19 $r = (3x^2)/(1 + 2x^2)$ agrees with $|x|$ at $x = -1, -\frac{1}{2}, 0, \frac{1}{2}, 1$. Show that

$$\frac{1}{12} \leq \max_{-1 \leq x \leq 1} \left| |x| - \frac{3x^2}{1 + 2x^2} \right| \leq \frac{1}{8}.$$

[*Hint:* It suffices to consider $0 \leq x \leq 1$. Is Exercise 5.18 applicable?]

5.20   Let J denote any subset of at least $m + 1$ distinct integers of $\{1, \ldots, k\}$. Let

$$\omega_J(x) = \prod_{j \in J} (x - x_j).$$

Show that, in place of (5.4.8), $p(x)$ can be determined by the more general expression

$$p(x) = \omega_J(x) \sum_{j \in J} \frac{f_j q(x_j)}{\omega_J'(x_j)(x - x_j)}.$$

BIBLIOGRAPHY

The bibliography that follows includes a selection of books on approximation theory containing material supplementary to this book. We shall try to guide the reader further with comments on some of them.

The books of Cheney, Davis, Lorentz, Rice and Meinardus contain much material in common with this book and treat many additional topics as well. Thus Davis [1] deals with approximation and interpolation in the complex domain. Lorentz [1] is mathematically more mature and studies ϵ-entropy and its applications. Meinardus [1] is focussed on uniform approximation. Cheney [1] is a textbook on a somewhat higher level than ours. Rice [1] is a two-volume text whose second volume emphasizes approximation by non-linear families.

Natanson [1] and Todd [1] are introductory works requiring not too much mathematical background and easy to read. Timan [1] is a veritable encyclopedia on its subject. It contains, for example, information on approximation of functions of several real variables, which is not easily found elsewhere.

Butzer [1], Garabedian [1] and Handscomb [1] are collections of papers on various topics. They provide a panoramic view of approximation theory and illuminate those aspects of the landscape that were being most energetically cultivated at the time they were published.

BIBLIOGRAPHY

ACHIESER, N. I.
1. *Theory of Approximation* (New York: Ungar, 1956). Translated from Russian.

AHLBERG, J. H., E. N. NILSON, and J. L. WALSH
1. *The Theory of Splines and Their Applications* (New York: Academic Press, 1967).

Approximation of Functions. See GARABEDIAN, H. (Ed.)

BARRODALE, I., and H. YOUNG
1. "Algorithms for best L_1 and L_∞ linear approximations on a discrete set," *Numer. Math.* 8 (1966), pp. 295–306.

BATEMAN MANUSCRIPT PROJECT
1. Bateman Manuscript Project: *Higher Transcendental Functions, Vol. II* (New York: McGraw-Hill, 1953).

BIRKHOFF, G.
1. "Local spline approximation by moments," *J. Math. Mech.* 16 (1967), pp. 987–990.

BIRKHOFF, G., and C. DEBOOR
1. "Piecewise polynomial interpolation and approximation," *The Approximation of Functions*, H. Garabedian (Ed.) (New York: Elsevier, 1965), pp. 164–190.

BOEHM, B.
1. "Functions whose best rational Chebyshev approximations are polynomials," *Numer. Math.* 6 (1964), pp. 235–242.

BUTZER, P. L., and J. KOREVAAR (Eds.)
1. *On Approximation Theory; Proceedings of the Conference at Oberwolfach*, Aug. 4–10, 1963 (Basel: Birkhäuser, 1964).

CHENEY, E. W.
1. *Introduction to Approximation Theory* (New York: McGraw-Hill, 1966).

CHENEY, E. W., and T. H. SOUTHARD
1. "A survey of methods for rational approximation, with particular reference to a new method based on a formula of Darboux," *SIAM Rev.* 5 (1963), pp. 219–231.

COURANT, R.
1. *Differential and Integral Calculus, Vols. 1, 2* (New York: Interscience, 1937).

143

DANTZIG, G. B.
1. *Linear Programming and Extensions* (Princeton, N.J.: Princeton Univ. Press, 1963).

DAVIS, P. J.
1. *Interpolation and Approximation* (Waltham, Mass.: Blaisdell, 1963).

DAVIS, P. J., and P. RABINOWITZ
1. *Numerical Integration* (Waltham, Mass.: Blaisdell, 1967).

EHLICH, H., and K. ZELLER
1. "Schwankung von Polynomen zwischen Gitterpunkten," *Math. Z. 86* (1964), pp. 41–44.

EISENHART, C.
1. "Boscovich and the combination of observations," *Roger Joseph Boscovich*, L. L. Whyte (Ed.) (New York: Fordham Univ. Press, 1961) Chapter 9, pp. 200–212.

ERDÖS, P.
1. "Problems and results on the theory of interpolation. II," *Acta Math. Acad. Sci. Hungar. 12* (1961), pp. 235–244.

FORSYTHE, G. E.
1. "Generation and use of orthogonal polynomials for data fitting with a digital computer," *J. Soc. Indust. Appl. Math. 5* (1957), pp. 74–88.

GÁL, I. S.
1. "On sequences of operations in complete vector spaces," *Amer. Math. Monthly 60* (1953), pp. 527–538.

GARABEDIAN, H. (Ed.)
1. *Approximation of Functions* (New York: Elsevier, 1965).

HANDSCOMB, D. C. (Ed.)
1. *Methods of Numerical Approximation* (Oxford: Pergamon Press, 1966).

HILDEBRAND, F. B.
1. *Introduction to Numerical Analysis* (New York: McGraw-Hill, 1956).

JACKSON, D.
1. *Theory of Approximation*, Amer. Math. Soc. Colloq. Publ., Vol. XI (Providence, R.I.: Amer. Math. Soc., 1930).

JACOBI, C. G. J.
1. "Über die Darstellung einer Reihe gegebener Werte durch eine gebrochene rationale Funktion," *Crelle J. für die Reine und Angew. Math. 30* (1846), pp. 127–156; reprint in *Gesammelte Werke, Vol. 3*, Weierstrass (Ed.) (Berlin: Georg Reimer, 1884), pp. 479–511.

KARLIN, S., and W. J. STUDDEN
1. *Tchebycheff Systems: With Applications in Analysis and Statistics* (New York: Interscience, 1966).

KOROVKIN, P. P.
  1. *Linear Operators and Approximation Theory* (Delhi: Hindustan, 1960).

KRIPKE, B. R.
  1. Best approximation with respect to nearby norms," *Numer. Math. 6* (1964), pp. 103–105.

KRIPKE, B. R., and T. J. RIVLIN
  1. "Approximation in the metric of $L^1(X, \mu)$," *Trans. Amer. Math. Soc. 119* (1965), pp. 101–122.

LA VALLÉE POUSSIN, C. J., DE
  1. *Leçons sur l'Approximation des Fonctions d'une Variable Réelle* (Paris: Gauthier-Villars, 1952).

LORENTZ, G. G.
  1. *Approximation of Functions* (New York: Holt, Rinehart and Winston, 1966).

LUTTMANN, F. W., and T. J. RIVLIN
  1. "Some numerical experiments in the theory of polynomial interpolation," *IBM J. Res. Develop. 9* (1965), pp. 187–191.

MARKOV, V.
  1. "Über Polynome, die in einem gegeben Intervalle möglichst wenig von Null abweichen," *Math. Ann. 77* (1916), pp. 213–258.

MEINARDUS, G.
  1. *Approximation of Functions: Theory and Numerical Methods* (New York: Springer, 1967).

Methods of Numerical Approximation. See HANDSCOMB, D. C. (Ed.)

Modern Computing Methods
  1. *Modern Computing Methods*, Notes on Applied Science, No. 16, N.P.L., 2nd ed. (London: H.M.S.O., 1961).

MOTZKIN, T. S., and J. L. WALSH
  1. "The least pth power polynomials on a finite point set," *Trans. Amer. Math. Soc. 83* (1956), pp. 371–396.

NATANSON, I. P.
  1. *Konstruktive Funktionentheorie* (Berlin: Akademie-Verlag, 1955); *Constructive Function Theory*, Vols. *I, II, III* (New York: Ungar, *I:* 1964, *II:* 1965, *III:* 1965). Translations from the Russian.

NEWMAN, D. J.
  1. "Rational approximation to $|x|$," *Michigan Math. J. 11* (1964), pp. 11–14.

On Approximation Theory. See BUTZER, P. L. and J. KOREVAAR (Eds.)

PASZKOWSKI, S.
　1. *On Approximation with Nodes*, Rozprawy Matematyczne, XIV (Warsaw: PWN, 1957).
　2. *The Theory of Uniform Approximation. I: Non-Asymptotic Theoretical Problems*, Rozprawy Matematyczne, XXVI (Warsaw: PWN, 1962).

POWELL, M. J. D.
　1. "On best L_2 spline approximations," T.P. 264, Math. Branch, A.E.R.E., Harwell, Berks, England (Nov. 1966).
　2. "On the maximum errors of polynomial approximations defined by interpolation and by least squares criteria," *Comput. J. 9* (1967), pp. 404–407.

RABINOWITZ, P.
　1. "Applications of linear programming to numerical analysis," *SIAM Rev. 10* (1968), pp. 121–159.

RALSTON, A.
　1. "Rational Chebyshev approximation by Remes' algorithms," *Numer. Math. 7* (1965), pp. 322–330.

REMEZ, E. YA.
　1. *General Computational Methods of Chebyshev Approximation. The Problems with Linear Real Parameters*, AEC-tr-4491 (U.S. Atomic Energy Commission, 1962). Translated from Russian into two volumes.

RICE, J. R.
　1. *The Approximation of Functions, Vols. 1, 2* (Reading, Mass.: Addison-Wesley, *1:* 1964, *2:* 1968).

RIVLIN, T. J.
　1. "Polynomials of best uniform approximation to certain rational functions," *Numer. Math. 4* (1962), pp. 345–349.

RIVLIN, T. J., and E. W. CHENEY
　1. "A comparison of uniform approximations on an interval and a finite subset thereof," *J. Soc. Indust. Appl. Math. Ser. B Numer. Anal. 3* (1966), pp. 311–320.

RUBIO, J. E.
　1. "A computational method for the approximation of functions in the space $L[0, 1]$," *J. Math. Anal. Appl. 19* (1967), pp. 56–66.

SCHOENBERG, I. J.
　1. "On spline functions," *Inequalities* (New York: Academic Press, 1967), pp. 255–291.

SHARMA, A., and A. MEIR
　1. "Degree of approximation of spline interpolation," *J. Math. Mech. 15* (1966), pp. 759–767.

STEFFENSEN, J. F.
　1. *Interpolation* (Baltimore, Md.: Williams and Wilkins, 1927).

STIEFEL, E.
1. "Note on Jordan elimination, linear programming, and Tchebycheff approximation," *Numer. Math.* 2 (1960), pp. 1–17.

SZEGÖ, G.
1. *Orthogonal Polynomials*, Amer. Math. Soc. Colloq. Publ., Vol. XXIII, rev. ed. (Providence, R.I.: Amer. Math. Soc., 1959).

TIMAN, A. F.
1. *Theory of Approximation of Functions of a Real Variable* (New York: Macmillan, 1963). Translated from Russian.

TODD, J.
1. *Introduction to the Constructive Theory of Functions* (New York: Academic Press, 1963).

TURÁN, P.
1. "On the approximation of piecewise analytic functions by rational functions," *Contemporary Problems in the Theory of Analytic Functions* (International Conference, Erevan, 1965) (Moscow: Izdat. "Nauka", 1966), pp. 296–300.

USOW, K. H.
1. "On L_1 approximation. I: Computation for continuous functions and continuous dependence; II: Computation for discrete functions and discretization effects," *J. Soc. Indust. Appl. Math. Ser. B Numer. Anal.* 4 (1967), pp. 70–88; pp. 233–244.

VALENTINE, F. A.
1. *Convex Sets* (New York: McGraw-Hill, 1964).

WERNER, H.
1. "Rationale Tschebyscheff-Approximation, Eigenwerttheorie und Differenzenrechnung," *Arch. Rational Mech. Anal.* 13 (1963), pp. 330–347; English translation by Fraser and Fraser, Dept. of Computer Science, London (Canada: Univ. of Western Ontario), 1965.
2. *Vorlesung über Approximationstheorie*, Lecture Notes in Mathematics, Vol. 14 (Berlin: Springer, 1966).
3. "Die Bedeutung der Normalität bei rationaler Tschebyscheff-Approximation," *Computing* 2 (1967), pp. 34–52.
4. "Ein Satz über diskrete Tschebyscheff-Approximation bei gebrochen linearen Funktionen," *Numer. Math.* 4 (1962), pp. 154–157.

ZUKHOVITSKIY, S. I., and L. I. AVDEYEVA
1. *Linear and Convex Programming* (Philadelphia, Pa.: Saunders, 1966). Translated from Russian.

ZYGMUND, A.
1. *Trigonometric Series, Vol. I* (Cambridge, England: Cambridge Univ. Press, 1959).

INDEX